THE KEY TO SUCCESSFUL DATA MIGRATION

PRE-MIGRATION ACTIVITIES

RAJENDER KUMAR

Trademarks

Screenshots

Website References

All website references were current at the time of publication.

For more information, contact: support@JambaAcademy.com.

Published by:
Jamba Academy

Printed in the United States of America

First Printing Edition, 2023

Found Typos & Broken Link

We apologize in advance for any typos or broken link that you may find in this book. We take pride in the quality of our content and strive to provide accurate and useful information to our readers. Please let us know where you found the typos and broken links (if any) so that we can fix them as soon as possible. Again, thank you very much in advance for bringing this to our attention and for your patience.

If you find any typos or broken links in this book, please feel free to email me.

support@JambaAcademy.com

Support

We would love to hear your thoughts and feedback! Could you please take a moment to write a review or share your thoughts on the book? Your feedback helps other readers discover the books and helps authors to improve their work. Thank you for your time and for sharing your thoughts with us!

If there is anything you want to discuss or you have a question about any topic of the book, you can always reach out to us, and we will try to help as much as we can.

support@JambaAcademy.com

Dedicated to all the professionals who strive for seamless and successful data migration, this book is a testament to your dedication and hard work.

May it serve as a valuable resource in your journey towards achieving excellence in the field of data migration.

DISCLAIMER

The information contained in this book is provided on an 'as is' basis, without warranty of any kind, either express or implied, including but not limited to the implied warranties of merchantability, fitness for a particular purpose, or non-infringement. The authors and publisher shall have no liability to any person or entity with respect to any loss or damage caused or alleged to be caused directly or indirectly by the information contained in this book. The information contained in this book is for general information purposes only and is not intended to be a comprehensive guide to all aspects of the topic covered. It is not intended to be used as a substitute for professional advice or services, and the authors and publisher strongly advise that you seek the advice of an expert before taking any action based on the information contained in this book.

ACKNOWLEDGMENTS

I would like to express my heartfelt gratitude to my colleagues, who provided valuable feedback and contributed to the development of the ideas presented in this book. In particular, I would like to thank Kalpita Dapkekar for her helpful suggestions and support.

I am also grateful to the editorial and production team at Jamba Academy for their efforts in bringing this book to fruition. Their professionalism and expertise were greatly appreciated.

I also want to thank my family and friends for their love and support during the writing process. Their encouragement and understanding meant the world to me.

Finally, I would like to acknowledge the many experts and thought leaders in the field of data science whose works have inspired and informed my own. This book is the culmination of my own experiences and learning, and I am grateful to the wider community for the knowledge and insights that have shaped my thinking.

This book is a product of many people's hard work and dedication, and I am grateful to all of those who played a role in its creation.

HOW TO USE THIS BOOK?

Here are some suggestions for making the most out of the book:

I. **Start with a brief introduction:** To grasp the main goal and organization of the book, start by reading the introduction (Preface). You will be better able to comprehend the context and flow of the information presented as a result.

II. **Read the chapters in order**: The chapters are arranged logically and built upon one another to provide readers a thorough comprehension of the subject. To properly understand the principles offered, it is advised to read the chapters in the order they presented.

III. **Put the examples and exercises to use**: To aid readers in understanding and putting the principles covered to use, the book provides examples, exercises, and case studies. Make sure to complete them in the order they are presented in the book.

IV. **Reference the supplementary resources**: To give readers more information and support, the book includes a number of resources, including websites, books, and online courses. Use these tools to help you learn more and stay up to date on what's going on in the field.

V. **Use the information**: Using the knowledge offered in the book in real-world situations is the greatest method to fully comprehend and remember it. In order to get practical experience and consolidate your comprehension, try to use the principles we've covered in your own work or on personal projects.

VI. **Review the chapter summary and questions**: The summary and questions at the end of each chapter are meant to help you review and assess your comprehension of the subject.

Before beginning the following chapter, make sure to go over them again.

VII. **Ask for assistance if necessary**: Don't be afraid to ask for assistance if you're having trouble grasping a concept or need further support. Join online forums, go to meetups, or look for a mentor to help you get beyond any challenges you may face.

You may make the most of this book and obtain a thorough understanding of data analysis and its applications in the real world by using these suggestions.

CONVENTIONS USED IN THIS BOOK

When learning a new programming language or tool, it can be overwhelming to understand the syntax and conventions used. In this book, we follow certain conventions to make it easier for the reader to follow along and understand the concepts being presented.

Italics

Throughout the book, we use italics to indicate a command used to install a library or package. For example, when we introduce the Keras library, we will italicize the command used to install it:

!pip install keras

Bold

We use bold text to indicate important terminology or concepts. For example, when introducing the concept of **neural networks**, we would bold this term in the text.

Handwriting Symbol

At times, we may use a handwriting symbol to indicate an important note or suggestion. For example, we may use the following symbol to indicate that a certain code snippet should be saved to a file for later use:

This is an important note or information.

CONTENTS

ABOUT THE AUTHOR

Rajender Kumar, an IT specialist and data scientist, has always been fascinated with how data can help shape insightful conclusions and decision making. has a wealth of experience in the field of data science, machine learning, and big data. With over a decade of experience, the author has worked with numerous large companies, including Walgreens Boot Alliance, Healthcare Retroactive Audits, NEBA, and Loxia technologies, to name a few.

Over the years, the Rajender has perfected his skills in data migration by working on projects requiring extensive planning, preparation, and execution. Due to a thorough knowledge of all of the complexities associated with migrating data, he has gained expertise in pre-migration activities which he now share with readers through this book.

Rajender's professional experience spans data analysis, engineering, and data science roles within various organizations as well as consulting to several on data strategy, architecture, and management strategies. He has an impressive track record in designing and implementing data solutions tailored specifically for each organization based on unique needs.

Rajender took part in various initiatives throughout his career, from building predictive models for established businesses to designing custom data pipelines for emerging startups. In each of these projects, his primary focus was not only technical solutions but also on working closely with clients to comprehend their needs from a business perspective and assist them in reaching their objectives.

Rajender Kumar is a passionate traveler and reader in addition to his professional pursuits. When not engaged in work-related pursuits, he can often be found reading or traveling to new places with the goal of immersing himself into local cultures.

Rajender strives to find new challenges and opportunities to learn and grow in both his professional and personal lives, be they technological advances in his industry or discovering new interests outside work. He values lifelong learning and development.

His passion for data and its importance stems from his belief that effective data management is critical to business success. He take data as one of an organization's most precious resources and must be handled and transferred with extreme care and precision. His dedication can be seen in his meticulous planning, preparation and execution for every project he works on.

In this book, he shares his expertise in pre-migration activities. He offers practical guidance for how to prepare for data migration projects by setting project scope and objectives and creating a communication plan; furthermore, he share best practices regarding risk mitigation, data profiling and mapping.

Readers of this book will gain from the author's extensive experience in data migration. They will gain valuable insight into the challenges and complexities involved with data migration projects as well as practical strategies and techniques for ensuring their successful execution. Whether you are an analyst, engineer, scientist or business leader involved with migration projects this book serves as an indispensable resource.

OTHER BOOK BY THE AUTHOR

You can get all the book from here:

https://linktr.ee/rajender03

- Mastering Data Analysis with Python: A Comprehensive Guide to NumPy, Pandas, and Matplotlib (ISBN: **978-1-960833-03-7**)
- Python Machine Learning: A Beginner's Guide to Scikit-Learn (ISBN: **978-1-960833-01-3**)
- The Bhagavad Gita: A Guide to Living with Purpose: A Search for Meaning in Modern Life
- Nugget of Wisdom from the Quran: Life lesson and teaching of the Holy Quran for a peaceful and happy life

WHO THIS BOOK IS FOR?

Data migration can be an intimidating, complex, and ever-evolving world to navigate without guidance and assistance from knowledgeable guides. That's why we wrote this book: to equip readers with all of the skills and information needed for successful data migration. But who exactly are its readers? Here is possible audience of the book:

Business Leaders: This book is tailored towards anyone overseeing a data migration project - be it CEO, CIO or IT manager - providing them with all of the insights required to make intelligent decisions regarding their data migration strategy.

Data Architects: Are You Responsible for Designing Data Architecture at Your Organization? This book is written specifically with you in mind! Here, we will review best practices in data modeling, mapping and integration as well as strategies to optimize storage and retrieval.

Data Scientists: Data scientists will find much to gain from reading this book, as we discuss techniques for data profiling, cleansing, quality assurance and management as well as ways to handle unstructured sources.

IT Professionals of All Levels: IT professionals of all levels will gain from the practical advice and technical knowledge provided in this book, whether you are a **database administrator, system administrator, software engineer or engineer** working on data migration projects. You will learn how to effectively plan and execute one successfully.

Consultants and Advisors: Consultants and advisors who collaborate with clients on data migration projects will find this book particularly helpful. It contains best practices for scoping projects, managing stakeholder expectations, and producing successful outcomes.

Students and Academics: Both students and academics in computer science, information systems or data science will find this book to be an indispensable source for comprehending how concepts they've studied in class can be implemented practically in everyday life.

This book is designed for anyone interested in data migration best practices and techniques, from business leaders, data architects, data scientists, IT professionals, consultants, advisors, students or academics - it will equip you with all of the knowledge you need to plan, execute, and deliver successful data migration projects. So if you want to take your data migration game to the next level - then this book is exactly what you need!

WHO THIS BOOK IS NOT FOR?

This book aims to give readers a deeper understanding of the pre-migration activities necessary for successful data migration, yet may not be suitable for every reader. Below are a few potential readers who might not benefit from reading it:

- People already familiar with data migration may not find this book to be as informative and beneficial.

- This book may not be ideal for readers seeking a technical or step-by-step instruction manual on data migration, as its focus lies more in pre-migration activities that often go overlooked in such projects.

- Anyone searching for a book focusing solely on the technical aspects of data migration, such as coding and scripting, may not find this book to be useful as an authority source.

- Readers looking for an easy or quick solution to data migration will not benefit from this book; successful data migration requires careful planning, execution and management that takes considerable dedication and hard work.

Overall, this book is intended for individuals engaged in data migration projects who wish to gain a deeper understanding of pre-migration activities which can greatly influence the success of their projects.

WHAT ARE THE REQUIREMENTS? (PRE-REQUISITES)

Before venturing into the intricacies of successful data migration, readers must satisfy certain prerequisites in order to gain a full grasp of its concepts presented herein.

First and foremost, readers should have a basic understanding of databases, data structures and data management principles.

Second, readers should possess experience in project management methodologies, including planning, implementation, and execution of complex projects. This will allow readers to comprehend the importance of pre-migration activities as well as careful planning and execution.

Thirdly, readers should become acquainted with various data migration tools and software and gain experience implementing and using them in their organization. This will allow readers to fully appreciate both benefits and limitations associated with different migration approaches while selecting an ideal solution for themselves.

Final note for readers should include an open mindset and willingness to adapt and learn new technologies and processes. Successful data migration involves more than technical expertise; it requires commitment to continuous improvement as well as openness to embrace change.

By fulfilling these pre-requisites, readers will gain the necessary skills and knowledge for successful implementation of pre-migration activities outlined in this book and ultimately a successful data migration.

PREFACE

As technology evolves, businesses and organizations face increasingly complex data management issues. Data migration - the process of moving information from one system or platform to another - has become more challenging as more data and applications come online. In this book we'll focus on critical pre-migration activities necessary for successful data migrations.

This book serves as a comprehensive guide for organizations and individuals who are planning a data migration. We will explore its purpose, different types of migration that organizations might experience and the key components for its successful execution - project planning, data profiling, cleansing and mapping.

Through real-life examples and case studies, this book will demonstrate key concepts and principles related to data migration. Furthermore, organizations can utilize various tools and software applications available for aiding their data migration process. This book is written with project managers, data analysts, database administrators and IT professionals in mind.

This book's purpose is to assist organizations successfully navigate the complexities of data migration by offering an in-depth guide for pre-migration activities. By adhering to its best practices and principles, organizations can ensure their data migration will go as smoothly as possible and enjoy its many advantages.

WHY SHOULD YOU READ THIS BOOK?

Are You Searching for an easy solution to Data Migration? Do you want to avoid costly errors and minimize downtime during the migration process? Here is your solution - "The Key to Successful Data Migration: Pre-Migration Activities."

This book aims to equip you with the knowledge and tools required for an efficient data migration project. Here are just a few reasons why you should read it:

- **Learn the Importance of Pre-migration Activities**: For any data migration project to succeed, careful pre-migration planning and preparation are vitally important. This book details all necessary steps that need to be taken prior to beginning migration in order to guarantee its success and limit risks.

- **Gain an understanding of common challenges and risks**: Data migration projects can be complex affairs with numerous risks attached. This book details common obstacles encountered during migration as well as strategies to combat any that arise during this process.

- **Discover Best Practices and Case Studies**: This book presents best practices and case studies from industry experts, showcasing effective data migration processes and techniques.

- **Discover tools and software solutions**: This book presents an overview of data migration tools and software solutions which can assist with the migration process, saving both time and money in the process.

This book provides project managers, IT professionals, and business executives who are charged with overseeing the migration of data from legacy systems to modern platforms, with invaluable advice for making it a smooth process. From planning and scoping through profiling mapping cleansing it covers it all - offering insight into tools available for data migration as well as practical tips on how best to utilize them effectively.

No matter if it's your first big data migration project or simply looking to gain a better understanding of data migration processes, this book serves as an essential resource to navigate through its complexities and ensure a successful result. Packed with practical insights and expert guidance from industry veterans as well as real-life examples - it should be required reading for anyone involved with data migration projects!

By reading "The Key to Successful Data Migration: Pre-Migration Activities," you will gain valuable insight into the pre-migration activities and best practices essential for successful data migration projects. No matter if you are an IT professional, data analyst, or business owner; "The Key to Successful Data Migration: Pre-Migration Activities" serves as an indispensable guide in making your data migration project a success.

Rajender Kumar

THE KEY TO SUCCESSFUL DATA MIGRATION

PRE-MIGRATION ACTIVITIES

01

1 INTRODUCTION TO DATA MIGRATION

In this chapter, we will delve into the world of data migration and explore the various facets of this complex process. We will begin by defining what data migration is and why it is important for businesses. We will discuss the evolution of data migration and its impact on business transformation, as well as the common reasons why organizations undergo data migration.

We will also take a closer look at the different types of data migration, including system migration, database migration, and application migration, among others. Each type of migration presents unique challenges that must be addressed to ensure a successful outcome.

Additionally, we will examine some of the key challenges that organizations face when undertaking a data migration project, such as data quality, data security, and compliance with regulatory requirements. We will explore some of the strategies and techniques that can be used to overcome these challenges and ensure a successful migration.

Finally, we will discuss the relationship between data migration and business transformation, and how a successful data migration can lay the foundation for a more agile, responsive, and data-driven organization. We will also touch on the role that data migration plays in digital transformation and the emergence of new technologies such as cloud computing, big data, and artificial intelligence.

By the end of this chapter, you will have a deeper understanding of what data migration entails and the various factors that contribute to its success. You will be better equipped to plan and execute a data migration project that meets your organization's needs and supports its business goals.

1.1 DEFINITION AND IMPORTANCE OF DATA MIGRATION

Data migration refers to the process of transferring data from one system or format to another while preserving its integrity and usability. It is a complex and multifaceted undertaking that involves a variety of technical, organizational, and strategic considerations.

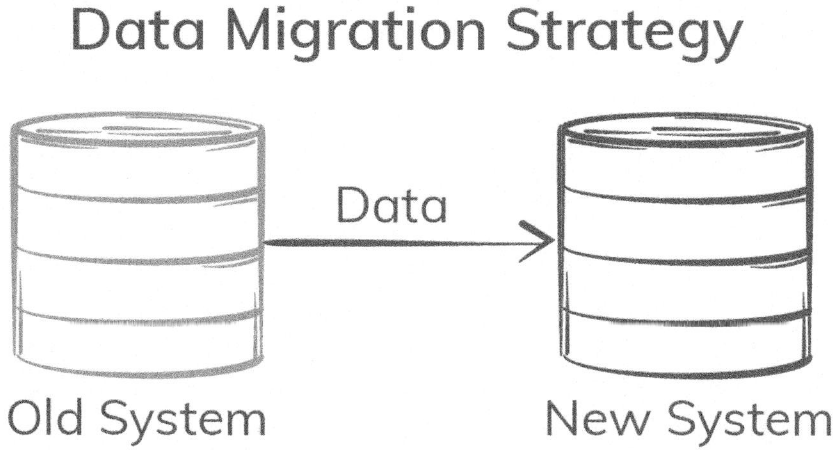

Data migration is at its core a process to guarantee data can be utilized effectively and efficiently for business operations and decision-making. This may involve migrating data from an outdated legacy system to a modern one, consolidating multiple databases, or moving it to the cloud. Ultimately, data migration strives to guarantee accuracy, completeness, consistency in all aspects of storage as well as access by those who require it.

Data migration is of critical importance. As businesses grow and technology progresses, the need for modern data infrastructure becomes even more pressing. Organizations must be able to rapidly transfer data between systems, guarantee data quality and security, as well as meet regulatory requirements.

Data migration can have a major effect on the success of a business. Improper or poorly executed data migration can lead to lost or incorrect information, as well as other problems that harm operations, customer relationships and revenue streams. Conversely, successful data migration enables organizations to take advantage of new technologies, gain insights from their data, and make informed decisions that promote growth and innovation.

Organizations seeking to successfully migrate data must approach the process strategically and with a detailed plan. This involves identifying source and target systems, mapping data elements, transforming it, assuring data quality, managing migration scripts and code, as well as addressing key challenges like data security, legal/ regulatory compliance, and stakeholder engagement.

Example:

A real-world example of data migration occurs when a large retail chain upgrades its point-of-sale (POS) system. The old POS was outdated and no longer met the business requirements, so they decided to implement a modern POS that provided real-time inventory management, detailed sales reporting, and customer analytics.

To accomplish this goal, the retail chain had to migrate their existing data from an old POS system to the new one. This included mapping fields, assuring data quality, and altering it according to the new system's needs. Furthermore, security during migration was essential in protecting sensitive customer information during this process.

1.2 THE EVOLUTION OF DATA MIGRATION AND ITS IMPACT ON BUSINESS

Data migration has been a remarkable journey that has dramatically altered how organizations store, access, and utilize their information. From paper-based records to modern cloud computing and big data analytics technologies, the methods and tools employed for data migration have continued to develop in line with changing business requirements and technological advancements.

In this section, we'll investigate the history of data migration and its effect on business, from manual data entry to today's advanced automation and machine learning technologies.

The Early Days of Data Migration

At the dawn of data migration, it was largely manual and paper-based. Organizations would store their information on paper records or punch cards and manually enter it into computer systems using keyboards or other input devices.

In the 1960s and '70s, mainframe computers revolutionized data storage and management. These powerful systems could store and process large amounts of information quickly, making them indispensable to many large organizations for business-critical functions like financial management, inventory control, and customer relationship management.

However, migrating data from one mainframe system to another was still largely manual, necessitating significant time and energy for mapping fields, transferring files, and assuring data quality.

The Rise of Relational Databases

Relational databases were first introduced in the 1980s, marking an important turning point in data migration. Relational databases offered a more structured and organized method for data storage and management, enabling organizations to store and access their information more efficiently and scalable.

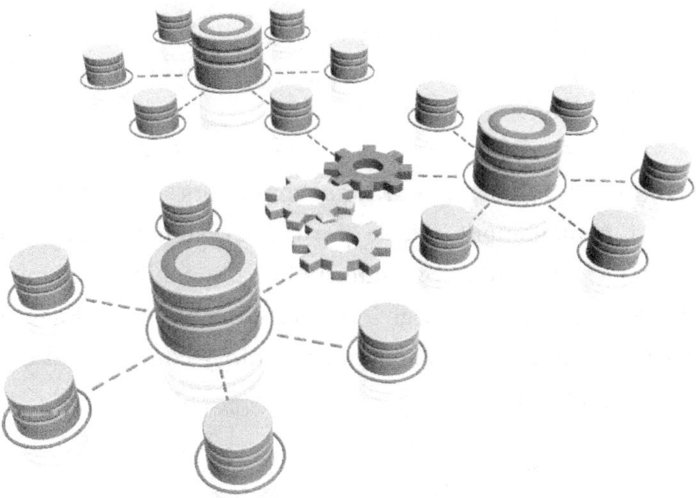

This new approach to data storage and management also created the platform for more advanced data migration techniques, such as **database migration** and **schema migration**. These procedures allowed organizations to move their data from one database system to another or from one schema to another while maintaining data integrity and usability.

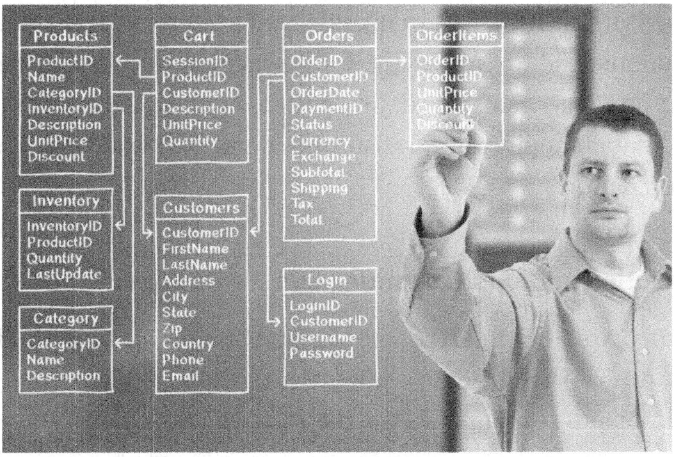

The Rise of Enterprise Resource Planning

In the 1990s and early 2000s, enterprise resource planning (ERP) systems revolutionized data migration. ERP programs offered a unified view of business operations, connecting all aspects from sales to manufacturing and supply chain management. With ERP systems, businesses could take an integrated approach to operations that was previously impossible.

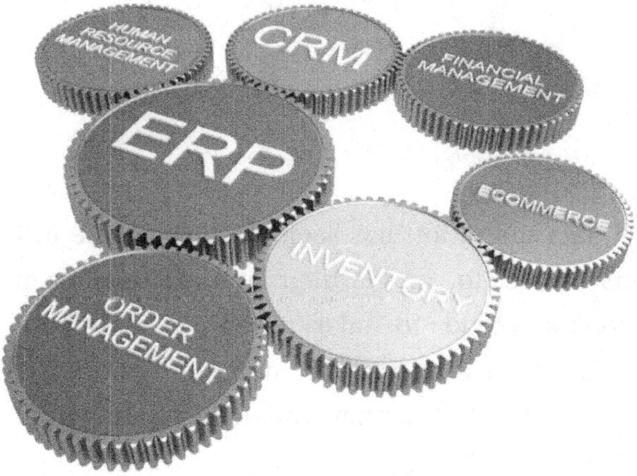

Implementing an ERP system necessitated extensive data migration, as organizations had to move their information from multiple legacy systems into one centralized, integrated one. This complex process involved data mapping and transformation as well as quality assurance checks on data.

The Evolution of Cloud Computing and Big Data

In recent years, the proliferation of cloud computing and big data has further transformed data migration. Cloud computing allows organizations to store and access their information more conveniently, while big data analytics provides new insights that can guide data-driven decision making.

Cloud-based data migration has seen a meteoric rise in popularity, as organizations strive to move their information from on-premise systems into the cloud to take advantage of cloud computing's advantages. This process necessitates new approaches such as cloud-to-cloud migration, hybrid migration, and data streaming.

Big data analytics has presented new challenges and opportunities for data migration, as organizations strive to gain value from large amounts of information from various sources. Data migration techniques such as **data lake migration** and **warehouse migration** have emerged to assist organizations with migrating their data onto big data platforms like Hadoop or Spark.

The Impact of Data Migration on Business

Data migration has had a transformative effect on businesses, enabling organizations to store, access, and utilize their information more efficiently and effectively. By migrating their data to more modern systems with greater flexibility and agility, organizations have gained new insights, improved decision-making processes, and enhanced customer experiences.

Data migration has been a critical element in business transformation, helping organizations transition away from legacy systems and towards more modern, agile ones that support their strategic objectives. By implementing new systems and processes, businesses can become more efficient, responsive, creative, and innovative - helping them stay ahead of the competition.

Data migration has had a tremendous effect on IT operations, allowing IT teams to streamline processes, reduce costs, and enhance data security and compliance. By adopting modern data migration techniques and tools, IT pros can automate many aspects of the migration process, decreasing errors and ensuring an efficient yet effective migration.

Data migration also presents unique challenges and risks, particularly around data quality and security. A poorly executed data migration can lead to loss, corruption, errors, which have a substantial effect on business operations as well as reputation. Furthermore, with the rise of big data and cloud computing comes greater responsibilities for organizations in terms of data security and privacy; organizations must implement robust security and compliance protocols in order to remain compliant.

1.3 COMMON REASONS FOR DATA MIGRATION

Data migration is a complex process that involves moving data from one system or format to another. Organizations may choose to conduct data migration for various reasons, such as upgrading outdated systems and consolidating multiple databases. In this section, we'll look at some common motivations behind data migration and their consequences for businesses.

Legacy System Upgrades

Data migration often begins as an effort to upgrade legacy systems, which are outdated technologies that no longer meet business demands. Upgrading these systems can provide organizations with new functionality, improved security and improved performance.

Upgrading legacy systems often necessitates data migration to transfer data from the old system to the new one. This process can be complex and time-consuming, necessitating meticulous planning and execution in order to guarantee accurate transfer of information securely.

One example of a legacy system upgrade is the case of a global retail company. They were using an outdated point-of-sale system that was slow and unreliable, leading to significant issues such as lost sales, unhappy customers, and reduced productivity.

To address these problems, the company decided to upgrade their point-of-sale system with a more modern one. This involved migrating all data from the old system over, including sales information, inventory data and customer records.

The migration process was intricate and necessitated meticulous planning and execution. To guarantee accurate data transfer, as well as minimal disruption to business operations during this transition period, the company had to take extra precautions.

To achieve this goal, the company collaborated with a team of data migration specialists who identified key data elements to be migrated and created an elaborate migration plan to guarantee everything ran smoothly.

System Consolidation

Organizations may also choose to conduct data migration in order to combine multiple systems into a single one. Doing so can streamline operations, cut costs, and enhance data management.

System consolidation necessitates data migration to move information from multiple systems into one unified one. This process can be challenging as organizations must guarantee that the data is accurately mapped and transformed for use within the new system.

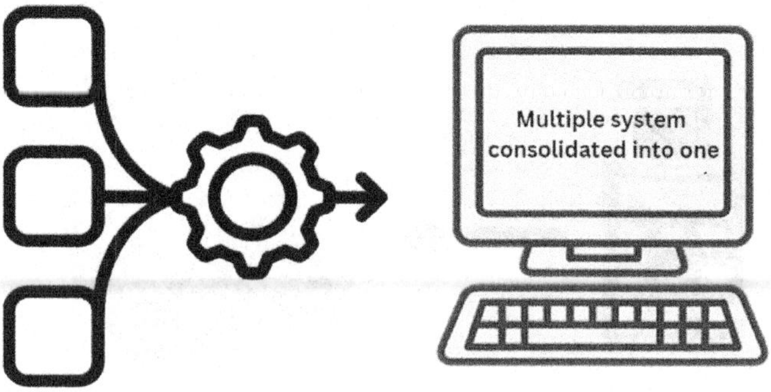

One example of such a scenario is a large retail chain that has multiple stores across different regions, each with their own inventory management system. To simplify operations and gain better insight into stock levels, they decide to consolidate these systems into one central inventory management system.

Cloud Migration

As more organizations move their data to the cloud, cloud migration has become a frequent cause of data relocation. This involves moving information from on-premise systems into cloud-based services like storage or applications hosted in the cloud.

Cloud migration is becoming more and more popular among businesses due to its flexibility, scalability, and cost-efficiency. Netflix serves as one example of cloud migration in action.

Netflix is a well-known streaming service that has revolutionized the entertainment industry. To guarantee high-quality streaming to millions of users around the world, Netflix needed to move its data into the cloud. Doing this enabled their servers to handle high traffic volumes associated with streaming video content.

Before moving to the cloud, Netflix relied on traditional data centers for storage and processing data. Unfortunately, this approach proved insufficient in the long run due to increasing demand for streaming services. As a result, they sought a more adaptable and scalable solution - hence why they turned towards cloud migration.

Netflix chose Amazon Web Services (AWS) for their cloud migration. AWS provided the necessary infrastructure and tools to move Netflix's data and applications into the cloud, with data migrating to S3 on AWS and application servers moving to Amazon Elastic Compute Cloud (EC2).

The advantages of cloud migration were abundant. Netflix could scale its infrastructure in real-time to meet user demand, providing faster and more dependable streaming services for viewers. Furthermore, they were able to reduce their infrastructure costs and focus on what they do best: providing entertainment.

This example emphasizes the significance of cloud migration for achieving business objectives. It illustrates how cloud migration can enhance a company's IT infrastructure's scalability, dependability, and cost-efficiency - allowing it to focus on core operational activities.

Cloud migration requires careful planning and execution to guarantee data is moved securely and efficiently. Organizations must also consider data sovereignty and compliance regulations when migrating data to the cloud.

Mergers and Acquisitions

Mergers and acquisitions are another common reason for data migration. When two organizations merge or one acquires another, they must integrate their information into a single system.

Data migration is necessary to transfer information from an acquired organization's system to that of the acquiring organization. This can be a challenging project, as organizations must guarantee that data is accurately mapped and transformed for use within their new platform.

Mergers and acquisitions (M&A) are often the catalyst for data migration, as the merging of two companies necessitates consolidation of information from multiple sources. A notable real-life example of M&A data migration is Microsoft's acquisition of LinkedIn.

In 2016, Microsoft acquired LinkedIn and began merging the data from two distinct companies. To accomplish this task, they took a two-step migration approach. The initial stage involved moving all LinkedIn data into Microsoft's Azure cloud service - an effort which involved transferring approximately 2.5 petabytes of information including user profiles, job postings, and other content.

The second stage involved migrating LinkedIn's application to Microsoft Azure, a process which took several months. Microsoft had to guarantee that LinkedIn users could continue access their accounts while also integrating LinkedIn functionality with other products and services offered by the company.

One of the challenges in this migration was ensuring the data was migrated securely, given LinkedIn's large amount of personal information. Microsoft used encryption to protect this information during transfer and also implemented additional security measures to prevent data breaches.

Data Center Consolidation

Data center consolidation involves moving data from multiple data centers into a centralized facility. This can reduce costs, enhance security, and streamline operations for organizations.

Data center consolidation necessitates data migration, moving information from multiple data centers into a centralized one. This task can be daunting as organizations must guarantee that all data is accurately mapped and transformed for use within the newly formed data center.

Data center consolidation can be seen in action with the U.S. Federal Government's Data Center Consolidation Initiative (DCCI), launched in 2010 to reduce the number of data centers operated by government and promote cloud computing. By 2015, their goal was to reduce these centers from over 2,000 down to less than 1,200.

Homeland Security

One agency that was successful in consolidating their data centers as part of the DCCI was the Department of Homeland Security (DHS). DHS had over 40 data centers scattered around the US that were expensive to run and maintain. By consolidating these facilities, they were able to reduce costs, improve security, and boost efficiency.

To achieve consolidation, the DHS utilized virtualization, cloud computing and physical consolidation. They virtualized their servers to reduce the need for physical servers, moved some applications to the cloud and consolidated remaining servers into fewer locations - this reduced their data center count from over 40 down to two locations, leading to significant cost savings.

Application Migration

Finally, organizations may perform data migration to move applications from one system to another. This can provide organizations with new functionality or improved performance. Application migration necessitates data movement from the old application into the new one.

One real-life example of application migration is the case of a large e-commerce company that needed to move its customer-facing applications into the cloud. Their existing on-premises infrastructure wasn't sufficient enough to handle the increased volume of customer transactions, leading to slow response times and frequent downtime.

To address these challenges, the company decided to move its applications onto a cloud-based platform. This migration involved several steps such as analyzing existing applications, recognizing their dependencies between them, and selecting an appropriate cloud platform for migration.

The company utilized automation tools for migrating applications, which helped reduce time and minimize errors. Furthermore, they conducted thorough testing before and after the migration to guarantee that all applications worked as expected.

1.4 TYPES OF DATA MIGRATION

Data migration is the process of moving data from one system or format to another. There are various types of data migration, each with its own challenges and considerations. In this section, we'll look into these different types of data migration and their consequences for businesses.

Storage Migration

Storage migration refers to the process of migrating data from one storage system to another. This could involve moving data from an on premise system into the cloud or from one cloud-based storage system to another.

In 2010, Microsoft announced the decommissioning of Hotmail email service and migration of its users' data to Outlook.com platform. After 26 years, Hotmail had become outdated and no longer met users' needs; as a result, Microsoft decided to retire Hotmail and migrate all user data over to Outlook.com which had a modern, responsive interface.

Microsoft orchestrated a migration process that moved millions of users' emails, contacts and calendars onto its new platform while guaranteeing security during the transition. To ensure an efficient transition, they utilized automated scripts, load-balancing mechanisms and redundant data centers as part of its suite of tools and technologies.

Microsoft was able to complete their migration of over 300 million Hotmail users into Outlook.com within six weeks, despite some initial difficulties such as slower-than-expected load times and missing emails. It marked one of the largest storage migrations in history with this feat.

Storage migration requires meticulous planning and execution to guarantee data is moved securely and efficiently. Organizations must consider factors like data sovereignty, privacy, and security when undertaking storage migration projects.

Database Migration

Database migration refers to the process of moving data from one database system to another. This could involve migrating from an on-premise system into the cloud or from one cloud-based database system to another.

One real-life **example** of database migration occurred when Verizon Wireless, a telecommunications company in the United States, transitioned their customer service platform from an on-premises Oracle database to Amazon Web Services (AWS) cloud database.

This migration was undertaken to enhance the performance and scalability of their customer service system, which was struggling to cope with an ever-increasing volume of customers.

The migration process consisted of several steps, such as identifying the data to be transferred, selecting an appropriate AWS database instance, creating a migration plan and testing it to guarantee accurate transfer without interruption to customer service systems.

Verizon Wireless reported significant improvements in system performance and scalability, as well as cost savings due to the use of cloud-based infrastructure. The successful migration served to underscore the significance of careful planning and testing when transitioning to a new database environment.

Organizations must guarantee that data is accurately mapped and transformed to fit into the new database system.

Application Migration

Application migration refers to the process of moving data between applications. This could involve migrating information from an on-premise application into the cloud, or from one cloud-based application to another.

One real world example of application migration is Nationwide Building Society, a financial institution headquartered in the UK. In 2017, they initiated a massive migration of their applications to the cloud with the goal of increasing agility, scalability and efficiency.

The migration process necessitated migrating over 450 applications to the cloud, including essential ones like online banking, mobile banking and payment systems. Furthermore, new technologies like containers and microservices architecture were adopted for this project.

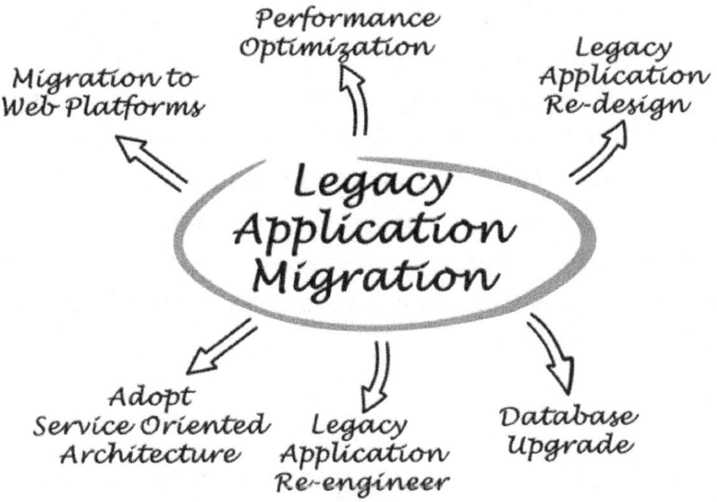

Nationwide faced several obstacles during their migration, such as legacy code and technical debt, complex application dependencies, and an abundance of manual processes. Furthermore, they had to guarantee that there would be no disruption in customer experience during the transition and all security and compliance requirements were fulfilled.

Nationwide took a gradual approach to migrating its applications, beginning with non-critical ones and gradually moving onto more critical ones. Furthermore, they invested in automation tools to simplify the migration process and reduce manual effort.

Nationwide's migration, which took two years to complete, brought about significant benefits for the company. They experienced increased agility and scalability, improved customer experience, cost savings through cloud technologies adoption. Furthermore, this success has allowed Nationwide to continue innovating and develop new applications and services for their users.

Operating System Migration

Operating System Migration is the process of moving an existing computer operating system from one hardware environment to another, such as from a physical server to a virtual machine or from an older server to a more modern one. It involves moving all relevant applications, data and configurations over to the new setup and making sure everything functions normally after migration is complete. The goal of an Operating System Migration is to minimize downtime while guaranteeing all users can access their applications and data without interruption.

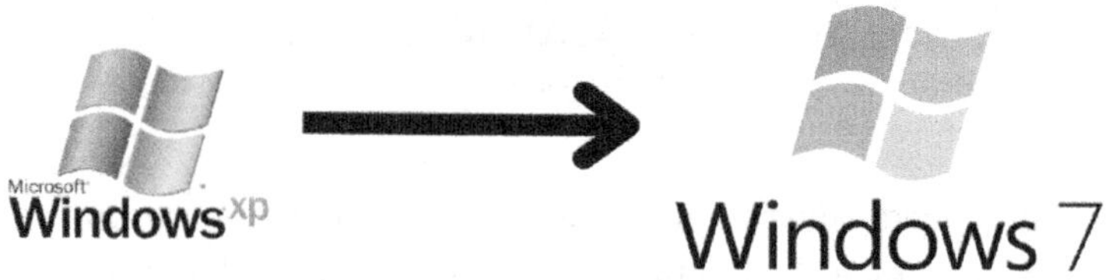

One successful case of operating system migration was New York City's Department of Finance in 2014. At that time, they were still running on an outdated Windows XP platform which wasn't supported by Microsoft anymore, posing a significant security risk to sensitive financial information for city residents.

To address this problem, the DOF launched an extensive OS migration project to transition to the more recent Windows 7 platform. This involved upgrading hardware, testing and updating software, as well as making sure critical business applications were compatible with the new platform. Furthermore, data was transferred from the old platform through both manual and automated methods.

During the migration project, The DOF faced several difficulties, such as extensive user training on the new platform and potential data loss during transit. Nevertheless, with careful planning, effective communication, and rigorous testing the project team was able to overcome these obstacles successfully.

The benefits of the OS migration project were immense for DOF and its users. The new platform provided enhanced security, faster system performance, and enhanced support for critical business applications. Furthermore, it enabled DOF to stay abreast of rapidly advancing technology and industry standards.

Platform Migration

Platform migration refers to the process of moving an application or system from one platform (such as hardware, operating system, database, programming language or development environment) to another in order to enhance performance, scalability reliability or maintainability of the system or meet changing business requirements. This can be a complex and risky undertaking that necessitates careful planning, testing and execution to guarantee that the system continues functioning normally after migration; it may also involve re-engineering or customizing the existing system so it works with the new platform.

One real-life example of platform migration is the shift of WhatsApp from Blackberry to more versatile platforms like iOS and Android. Initially designed for use on Blackberry smartphones, WhatsApp became immensely popular among its users. But as Blackberry's popularity decreased and iOS and Android became more prevalent, its developers decided to migrate their service over to these two platforms in order to reach a wider audience.

Migrating the WhatsApp application to iOS and Android platforms required porting it over and ensuring its smooth performance on these systems. Furthermore, the WhatsApp team had to guarantee the consistency of user interface and experience across different platforms, as well as making sure the application worked correctly with different hardware and software configurations across devices.

WhatsApp had to ensure the migration process didn't disrupt user experience or lead to data loss. They provided clear instructions on how to migrate their accounts onto new platforms, plus support if users experienced issues during the transition process.

Cloud Migration

Cloud migration is the process of migrating an organization's data, applications, and other business elements to a cloud-based infrastructure. This involves moving on-premises digital assets into either public, private, or hybrid clouds. The primary aim is to improve scalability, reduce IT expenses, and boost business agility by taking advantage of cloud computing's flexibility and advantages.

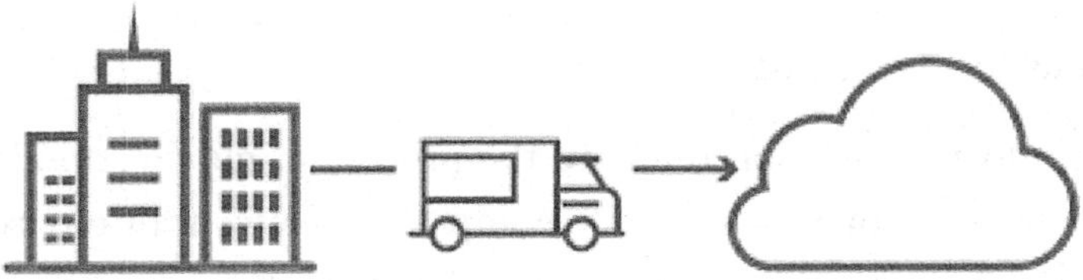

Marriott International, one of the world's largest hotel chains, undertook an ambitious cloud migration project in 2017. They had a decentralized IT infrastructure with numerous applications and systems spread across multiple data centers which presented challenges with data management, scalability, and security. To address these concerns, Marriott decided to move their entire IT infrastructure into the cloud.

Marriott's cloud migration project involved moving all of its applications, data, and IT systems into a private cloud environment. To design and execute this strategy, Marriott worked with an external service provider. Each application was assessed for compatibility with the cloud environment; those that weren't suitable were either retired or re-engineered for it.

Marriott also had to address data security concerns during the migration process. To guarantee its data was protected in the cloud environment, Marriott implemented advanced security measures like encryption and multi-factor authentication. The migration project took over a year to complete and involved over 5,000 applications and 250 databases that needed moving.

After the migration was complete, Marriott experienced significant improvements in their IT infrastructure, such as increased scalability, better data management and enhanced security. Furthermore, they were able to reduce their IT costs by over 20% as a result of the migration.

Cross-Platform Migration

Cross-platform migration refers to the process of moving applications, data, or other resources from one computing platform or environment to another. This may involve migrating from one type of hardware (like a mainframe to cloud-based environment) or software platforms (from Windows-based system to Linux-based one) from one environment to another.

Cross-platform migration can be a complex and time-consuming task, as different platforms often utilize distinct programming languages, data formats, and operating systems. To guarantee successful migration of resources that function properly in their new environment requires significant planning, testing, and optimization.

One real-life example of cross-platform migration is the transition of a banking institution's legacy systems from mainframe to distributed.

In the early 2000s, a major bank in America had an antiquated mainframe system that stored and managed customer information, transactions, and other vital data. This antiquated system couldn't keep up with growing demands from customers, leaving its IT team struggling to maintain it.

To address these challenges, the bank chose to migrate its legacy mainframe system to a distributed one. This involved migrating all data, applications and infrastructure from one platform that was unsuitable for their needs. To plan and execute this transition, their IT team worked alongside experts in migration techniques.

The migration process consisted of several stages, such as data extraction, cleansing, validation and migration. Furthermore, the IT team had to guarantee that the new system could integrate with other applications and systems used by the bank.

The migration was a success, providing the bank with an efficient and dependable platform to manage customer data and transactions. This enabled them to enhance customer service levels as well as boost operational efficiency - helping them remain competitive in an ever-evolving market.

Data migration presents organizations with a range of challenges and considerations, each having its own particular challenges. By understanding the different types of data migration, organizations can better plan and execute their projects efficiently while guaranteeing accuracy in data transmission to support business objectives.

1.5 KEY CHALLENGES IN DATA MIGRATION

Data migration is an intricate process that involves moving data from one system or format to another. While data migration can bring many benefits for organizations, such as increased efficiency and improved data quality, it also presents several challenges. In this section we'll look at the key challenges in data migration and discuss strategies on how organizations can successfully tackle them.

Data Quality

Data quality is a paramount element when migrating data. If the source data being migrated is of poor quality, the resulting information in the new system will also be of subpar quality, potentially leading to serious business issues.

Organizations must conduct a comprehensive data quality analysis before migrating their data, in order to address this challenge. This involves identifying and correcting any data quality issues such as missing information, duplicate records, or inconsistent data.

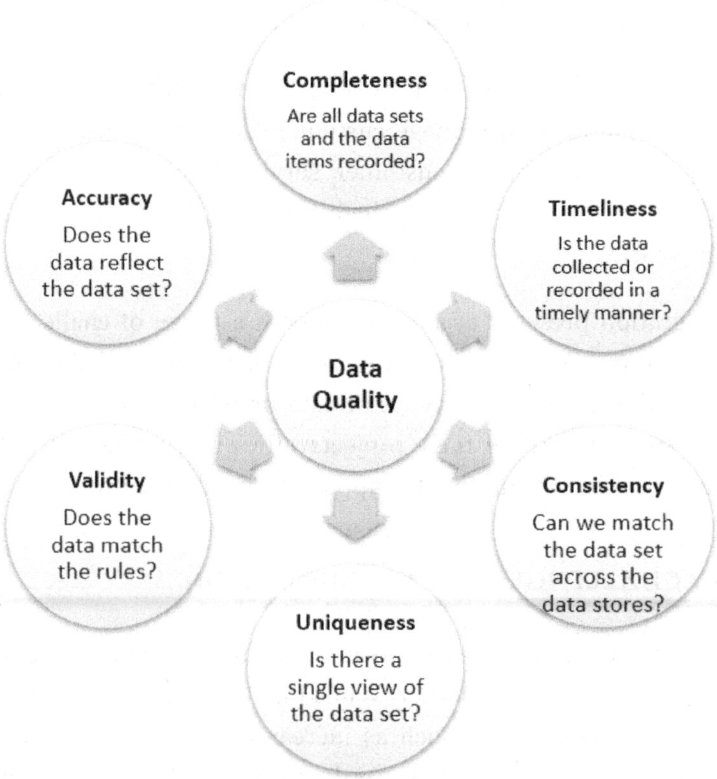

Data Mapping and Transformation

Data mapping and transformation involve mapping data fields from an old system to the new one, then altering it according to its data structure.

This process can be especially tedious for organizations with large, disparate data sets. To overcome this obstacle, organizations should create an extensive data mapping and transformation plan that includes detailed data mappings, transformation rules, and validation procedures.

Data Security

Data security is a paramount issue when migrating data. Organizations must guarantee that sensitive information is safeguarded against unauthorized access and modification during the transfer process.

To address this challenge, organizations should implement robust data security measures such as data encryption, access controls and data masking. Furthermore, they should conduct comprehensive security assessments to detect and address any security vulnerabilities.

System Downtime

Data migration often necessitates system downtime, which can be disruptive to business operations. To reduce the impact of system downtime, organizations should carefully plan their migration process and schedule it during periods of low activity.

Before migrating data, they should conduct thorough testing and validation of the new system to reduce the chance of system downtime.

Cost

Data migration can be costly, particularly for organizations with large and complex data sets. Costs may include hardware and software purchases, migration tool expenses, as well as consulting fees.

To address this challenge, organizations must assess their data migration needs and create a comprehensive migration plan with detailed cost estimates. Furthermore, they might look into using cloud-based migration tools and services in order to reduce expenses.

Data Governance

Data governance is an essential issue when migrating data. Organizations must guarantee that data migration follows data governance policies and regulations to guarantee it remains compliant.

Organizations should create a comprehensive data governance plan with detailed policies and procedures for data migration. They also need to conduct thorough compliance assessments in order to guarantee that data migration adheres to applicable regulations.

To overcome all of the above challenges, organizations must create a comprehensive migration plan that includes detailed data mapping and transformation rules, extensive testing and validation, as well as robust data security measures. They should also carefully assess their data migration needs and create an extensive cost estimate in order to anticipate any financial repercussions associated with data migration.

1.6 THE RELATIONSHIP BETWEEN DATA MIGRATION AND BUSINESS TRANSFORMATION

Data migration and business transformation are closely related concepts. Data migration involves moving data from one system or format to another, while business transformation refers to the process of fundamentally altering how a business operates.

Business transformation is a comprehensive process that involves altering how a business operates, usually in response to shifting market conditions, customer demands or technological advances. This may necessitate changes to processes, organizational structures or the adoption of new technologies.

Data migration is a crucial aspect of business transformation. By enabling data movement from one system to another, data migration encourages the adoption of new technologies and simplification of processes. Furthermore, it fosters the growth of new business models and generates additional sources of income.

Data migration can be an essential tool in supporting business transformation initiatives, providing several key benefits:

Data Migration as a Tool for Adopting New Technologies

Data migration allows the adoption of new technologies by transferring data from outdated systems to new ones. This opens the door for the implementation of innovative business models, the creation of new revenue streams, and product/service development that would not otherwise be possible.

For instance, a business transitioning from an on-premises system to the cloud may need to transfer its data there. Doing so allows them to utilize emerging cloud-based technologies like data analytics or artificial intelligence for improved operations and enhanced customer service.

Data Migration to Support Business Transformation

Data migration can support business transformation by streamlining processes. By migrating data to a new system, organizations are able to eliminate redundant steps and enhance their operations' efficiency.

For instance, businesses migrating their data to a new customer relationship management (CRM) system can eliminate manual data entry and enhance customer accuracy. This leads to faster response times and superior service levels, ultimately improving the organization's profitability.

Supporting Organizational Change

Data migration can facilitate organizational transformation by allowing organizations to adopt new business models or structures. By migrating data to a new system, businesses are better able to align their data with their strategic goals and foster new working styles.

For instance, businesses moving toward a customer-centric model may need to migrate their data into a new CRM system. Doing so allows the organization to better track customer interactions and preferences, leading to improved customer satisfaction scores.

Enabling Data-Driven Decision Making

Data migration can support business transformation by empowering data-driven decision making. By moving their information to a new system, organizations are able to better analyze their information and gain new insights into their operations.

For instance, businesses migrating their data to a new data analytics platform may be able to uncover emerging trends or patterns within their records. This helps facilitate informed decision making and ultimately boosts the organization's profitability.

Data migration and business transformation are inextricably linked. By facilitating data movement from one system to another, data migration can facilitate new technologies adoption, streamline operational processes, promote organizational change and enable data-driven decision making.

To maximize the advantages of data migration for business transformation, organizations must approach data migration strategically with an understanding of their business goals and how data plays a role in achieving them.

1.7 SUMMARY

- Data migration is a vital process that involves moving data from one system or format to another.
- It's necessary for organizations that need to move their information into a new platform in order to maintain business continuity and fuel growth.
- Data migration can be a complex and time-consuming task, necessitating careful planning, execution and testing to guarantee data is transferred accurately and securely.
- Prior to migration, key activities include identifying the data to be migrated, performing a data quality analysis and creating an exhaustive migration plan.
- Post-migration tasks involve data validation, testing and user acceptance testing.
- Organizations should prioritize data governance and compliance to guarantee that data migration is conducted in line with applicable regulations and policies.

- Achieving success with a data migration project necessitates meticulous planning, execution and ongoing support; additionally, strong data management practices guarantee accurate and secure transfer of the migrated information.

1.8 TEST YOUR KNOWLEDGE

I. What is data migration?

a) The process of encrypting data
b) The process of transferring data from one system or format to another
c) The process of creating new data
d) The process of deleting data

II. Why is data migration important?

a) It allows organizations to move their data to a new system and support business continuity.
b) It helps organizations to reduce their data storage costs.
c) It is a legal requirement for all businesses.
d) It is not important.

III. What are some key pre-migration activities?

a) Identifying the data to be migrated
b) Conducting a data quality analysis
c) Developing a comprehensive migration plan
d) All of the above

IV. What are some key post-migration activities?

a) Data validation
b) Testing
c) User acceptance testing
d) All of the above

V. What is a key challenge in data migration?

a) Data quality
b) Data security
c) Cost
d) All of the above

VI. Why is data governance important in data migration?

a) To ensure that data is migrated accurately and securely
b) To ensure compliance with applicable regulations and policies
c) To minimize the risk of data loss or corruption
d) All of the above

VII. What is the relationship between data migration and business transformation?

a) Data migration enables the adoption of new technologies
b) Data migration supports the streamlining of business processes
c) Data migration enables the adoption of new business models
d) All of the above

VIII. What is a key consideration for organizations undertaking data migration?

a) Careful planning and execution
b) Budget constraints
c) The use of the latest data migration tools and technologies
d) None of the above

IX. What is the role of data mapping and transformation in data migration?

a) To map data fields from the old system to the new system
b) To transform the data to fit the new system's data structure
c) To ensure data is migrated accurately and securely
d) All of the above

X. What is a key post-migration activity that organizations must undertake?

a) Data validation
b) Testing
c) User acceptance testing
d) Ongoing support and maintenance

1.9 ANSWERS

I. Answer b) The process of transferring data from one system or format to another
II. Answer a) It allows organizations to move their data to a new system and support business continuity.
III. Answer d) All of the above
IV. Answer d) All of the above
V. Answer d) All of the above
VI. Answer d) All of the above
VII. Answer d) All of the above
VIII. Answer a) Careful planning and execution
IX. Answer d) All of the above
X. Answer d) Ongoing support and maintenance

02

2 INTRODUCTION TO PRE-MIGRATION ACTIVITIES

Before beginning the actual data migration process, it is essential to have an extensive understanding of the pre-migration activities that must take place. These tasks include assessing migration needs, identifying data to be migrated, conducting a data quality analysis, creating an extensive migration plan and setting up data governance and compliance measures.

This chapter will define and analyze the significance of pre-migration activities, as well as identify some common difficulties organizations may experience during this stage. We'll also look at how data migration and pre-migration work together, emphasizing their connection and explaining how effective planning and execution of pre-migration tasks can set the stage for a successful data migration. By understanding pre-migration tasks' critical role in successful data migration projects and taking measures to ensure minimal disruptions to business operations, organizations can guarantee their projects run smoothly and successfully.

2.1 DEFINITION OF PRE-MIGRATION ACTIVITIES

Pre-migration activities refer to the planning and preparation steps taken before data migration begins. They guarantee a successful migration with accurate, complete, and high-quality migrated data. Pre-migration tasks involve assessing needs for data migration, identifying what data needs to be moved, conducting an analysis on its quality, creating an extensive migration plan, as well as setting up data governance and compliance measures.

Assessing Data Migration Needs

The initial step in the pre-migration process is to assess data migration needs. This involves identifying both business objectives driving the migration and specific data that needs to be moved. Organizations must carefully examine current and future business requirements, as well as which data assets are essential for operations. Collaboration among stakeholders such as business leaders, data owners, and IT personnel allows organizations to gain a comprehensive understanding of these demands and set priorities accordingly.

Identification of Data to be Migrated

Once the data migration requirements have been identified, the next step is to pinpoint exactly which data needs to be migrated. To do this, organizations must conduct a comprehensive inventory of all assets and decide which files need to be moved over. It is essential that organizations possess an in-depth knowledge of the structure, format and quality of the migrated information in order for the process to run efficiently. Furthermore, organizations should take into account any dependencies or relationships between datasets which must be preserved during migration.

Conducting Data Quality Analysis

Before migrating data to a new system, it is essential to guarantee the data is accurate, complete, and of high quality. This requires conducting a data quality analysis which involves identifying any errors or inconsistencies in the data and creating an action plan for resolving them - this may involve data cleansing, standardization, or other data quality improvement measures. Data quality analysis helps anticipate potential issues during migration processes and guarantees migrated data meets required quality standards.

Developing a Comprehensive Migration Plan

Once the data to be migrated has been identified and any quality issues addressed, organizations must create an extensive migration plan. This should include an organized timeline for the transfer, as well as resources needed for execution - personnel, technology and other - needed to finish the move. Furthermore, such an extensive plan must also include backup and recovery plans to guarantee no data is lost during transit.

Establishing Data Governance and Compliance Measures

Organizations must implement data governance and compliance measures to guarantee that their migration follows applicable regulations and policies. This may involve developing a data governance framework, implementing data security and privacy safeguards, as well as ensuring data migration maintains data integrity and security. Effective data governance and compliance protocols guarantee protected migrated information while remaining compliant with legal obligations.

2.2 IMPORTANCE OF PRE-MIGRATION ACTIVITIES

Pre-migration activities are essential for the successful completion of a data migration process. This section explores the primary reasons why pre-migration is so important and how it helps organizations meet their business objectives.

Minimizing Risks of Data Loss or Corruption

- Pre-migration activities help minimize these risks by minimizing exposure to potentially damaging scenarios.
- Data loss or corruption can have devastating effects for an organization.
- Pre-migration activities like data quality analysis help detect potential issues and guarantee the migrated data meets required quality requirements.

Assuring Accurate and Secure Data Migration

- Careful planning and execution of pre-migration activities can guarantee accurate, secure data migration.
- A comprehensive migration plan helps maintain data integrity and security throughout the process.

Streamlining the Migration Process

- Simplifying the migration process has never been simpler - simply start with proper planning!
- Pre-migration activities help streamline the migration process, minimizing its impact on business operations.
- By identifying specific data to be migrated, organizations can reduce the time required for this step.

Compliance with regulations and policies

- Neglecting to adhere to regulations and policies can have devastating results for an organization.
- Pre-migration activities, such as data governance and compliance measures, guarantee adherence to applicable rules and policies.

Maximizing Value from Migrated Data

- Migrated data should be maximized for maximum value creation through data processing operations.
- Pre-migration activities enable organizations to identify and prioritize critical data assets, maximizing the value from migrated information.
- Assessing data migration needs and conducting a data quality analysis are essential in this process.

2.3 KEY CHALLENGES IN PRE-MIGRATION ACTIVITIES

Data migration is a complex process with numerous obstacles that must be overcome for successful data transfer from one system to another. Pre-migration activities are essential components of this effort and present organizations with several unique difficulties they must conquer for the migration to take place smoothly. In this section, we'll address some key difficulties faced during pre-migration activities and how they can be overcome.

Data Quality and Consistency

A major challenge in pre-migration activities is guaranteeing the quality and consistency of data being migrated. Organizations must guarantee it's accurate, complete, and consistent across all systems; this necessitates conducting an exhaustive data quality analysis to identify any issues before migration begins.

Data Mapping and Transformation

Data mapping and transformation can present a major challenge during pre-migration activities. Organizations must guarantee that the data migrated is accurately mapped to the new system and transformed into the correct format, necessitating careful planning and execution of these operations for accurate results. This ensures the migrated information remains usable and accurate after migration.

Data Volume and Complexity

One of the biggest obstacles during pre-migration activities is dealing with the sheer volume and complexity of data being migrated. Organizations must ensure they have sufficient resources and tools for management and processing this information, which requires a deep comprehension of what needs to be migrated as well as being able to handle large datasets efficiently.

Data Governance and Compliance

Adhering to data governance and regulatory requirements can present a challenge during pre-migration activities. Organizations must make sure their migration process complies with all relevant regulations, policies (including data privacy/security standards), as well as having an in-depth understanding of the regulatory landscape in order to develop suitable data governance and compliance measures that adhere to it.

Resource Allocation

Resource allocation can be a significant challenge during pre-migration activities. Organizations must allocate the necessary resources, such as personnel, tools and equipment, to guarantee successful execution of pre-migration tasks. To do this effectively requires careful planning and coordination in order to guarantee those resources are available when needed.

2.4 PRE-MIGRATION ACTIVITIES AND RISK MITIGATION

Data migration is a complex and risky process if not managed carefully. It involves moving data from one system or environment to another, which could result in data loss, corruption or security breaches if done incorrectly. That's why having an organized strategy and performing pre-migration activities are so important; they help mitigate the risks associated with data migration. In this section we'll discuss their importance as well as how they can aid with risk reduction strategies.

Pre-Migration Activities

Pre-migration activities are essential for ensuring an efficient data migration process. These steps should include:

1. **Planning and Strategy Development:** It is essential to craft a comprehensive plan and strategy for data migration. This should encompass the scope, timeline, budget, and resources necessary for the transition process.

2. **Data Analysis and Profiling**: Data analysis and profiling are essential to comprehend the structure and format of data, helping identify any inconsistencies or issues that need to be resolved before migrating it.

3. **Data Cleaning and Transformation**: Data cleansing and transformation involves eliminating duplicates, resolving inconsistencies, and converting data into a standard format that is compatible with the target system.

4. **Backup and Disaster Recovery Planning**: Backup and disaster recovery planning guarantee that data is safeguarded during the migration process in case of any unanticipated events.

5. **Testing and Validation**: Validating the migrated data ensures it is accurate, complete, and in the desired format. Furthermore, this

guarantees compatibility with the target system so end-users can access and utilize it effectively.

Risk Mitigation

Pre-migration activities can help mitigate the risks associated with data migration. These risks may include:

1. **Data Loss or Corruption**: Data loss or corruption can occur due to technical problems, human error, or system malfunction during the migration process. Pre-migration activities like data analysis, backup planning, and disaster recovery planning help in minimizing the risk of data loss or corruption.

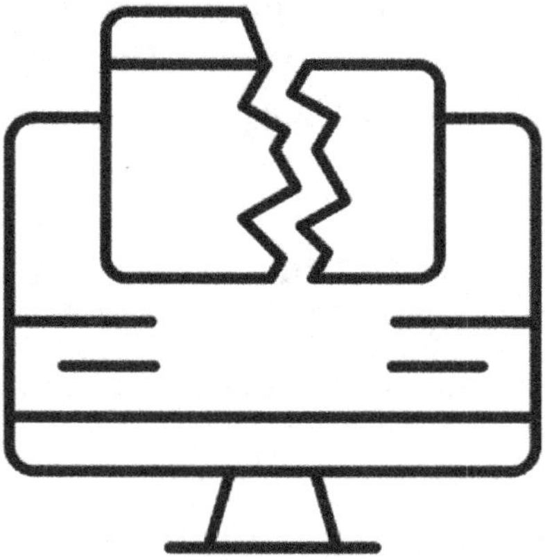

2. **Security Breaches**: Security breaches can occur during data migration due to unauthorized access, data leakage or hacking. Pre-migration activities like data analysis and profiling, testing and validation help in recognizing and addressing security issues prior to the migration process.

3. **System Downtime**: System downtime can occur during the migration process, impacting business operations and productivity. Pre-migration activities such as planning, strategy development, testing, and validation help minimize system downtime to guarantee that the migration is completed within its designated timeline.

2.5 SUMMARY

- Pre-migration activities are essential in guaranteeing a successful data migration process.
- These tasks include data quality analysis, mapping, and transformation.

- Pre-migration helps identify potential issues and create an exhaustive migration plan.
- Data quality is a critical aspect of pre-migration activities and ensures the accuracy and consistency of the data being migrated.
- Data mapping is another essential pre-migration activity that allows organizations to map their existing system's data into the new one.
- Risk mitigation is an essential pre-migration activity that helps organizations guarantee the success of the migration process.
- Pre-migration activities also enable organizations to create an achievable timeframe for migration and ensure compliance with applicable regulations and policies.

2.6 TEST YOUR KNOWLEDGE

I. Which of the following is NOT an example of a pre-migration activity?

a) Data quality analysis
b) Data mapping
c) Data migration
d) Risk mitigation

II. What is the importance of pre-migration activities?

a) To ensure the successful transfer of data from one system to another
b) To identify potential issues and challenges
c) To develop a comprehensive migration plan
d) All of the above

III. What is the purpose of data quality analysis in pre-migration activities?

a) To ensure that the data being migrated is accurate and complete
b) To map the data being migrated to the new system
c) To identify and mitigate risks
d) To develop a realistic timeframe for the migration process

IV. What is the purpose of data mapping in pre-migration activities?

a) To ensure compliance with applicable regulations and policies
b) To develop a comprehensive migration plan
c) To map the data being migrated to the new system
d) To identify potential issues and challenges

V. Which of the following is a key challenge in pre-migration activities?

a) Resource allocation
b) Data governance and compliance
c) Data mapping and transformation
d) All of the above

VI. What is the purpose of risk mitigation in pre-migration activities?

a) To ensure that the data being migrated is accurate and complete
b) To map the data being migrated to the new system
c) To identify and mitigate risks
d) To develop a realistic timeframe for the migration process

VII. What is the purpose of developing a realistic timeframe for the migration process?

a) To ensure that the data being migrated is accurate and complete
b) To map the data being migrated to the new system
c) To identify and mitigate risks
d) To minimize the impact on business operations

VIII. Which of the following is a critical aspect of pre-migration activities?

a) Compliance with applicable regulations and policies
b) Developing a comprehensive migration plan
c) Ensuring the quality and consistency of the data being migrated
d) All of the above

IX. Which of the following is NOT a pre-migration activity?

a) Data quality analysis
b) Data governance and compliance
c) Data transformation
d) Data migration

X. Which of the following is a potential challenge in pre-migration activities related to data quality?

a) Resource allocation
b) Data governance and compliance
c) Data mapping and transformation
d) Incomplete or inaccurate data

2.7 ANSWERS

I. Answer: c) Data migration is not a pre-migration activity but rather the end goal of the migration process.
II. Answer: d) All of the above.
III. Answer: a) To ensure that the data being migrated is accurate and complete.
IV. Answer: c) To map the data being migrated to the new system.
V. Answer: d) All of the above.
VI. Answer: c) To identify and mitigate risks.
VII. Answer: d) To minimize the impact on business operations.
VIII. Answer: d) All of the above.
IX. Answer: d) Data migration is not a pre-migration activity but rather the end goal of the migration process.
X. Answer: d) Incomplete or inaccurate data.

03

3 ESTABLISHING PROJECT SCOPE AND OBJECTIVES

As the planning phase of data migration begins, establishing project scope and objectives is an essential first step. Delineating these boundaries involves outlining both the intended outcomes as well as setting boundaries for what the migration project will entail. This chapter will emphasize the significance of clearly outlining project scope and objectives, as well as provide an overview of key elements involved. Additionally, this chapter will address identifying stakeholders and their roles in the process, as well as best practices for defining project scope and objectives. At this stage, challenges in the data migration process must also be identified and addressed; solving them is essential to create a clear project scope and objectives. By clearly outlining these details, organizations can guarantee that their data migration process is successful and produces the desired outcomes.

3.1 DEFINING PROJECT SCOPE

Definition of project scope is an essential step in any successful data migration project. It defines the boundaries and direction for the migration process. In this section, we'll explain why defining scope is so important and identify some key components involved; additionally, we'll provide a practical use case to illustrate these concepts.

What is Project Scope?

Project scope refers to the objectives, deliverables, and requirements for a data migration project. It specifies what data will be migrated, the systems involved in the migration, as well as an approximate timeline. A clearly defined project scope ensures all parties involved in the migration process are on board and working towards similar goals.

Why Defining Project Scope Is Important

Establishing project scope is critical in any successful data migration effort.

Determining project scope is essential for several reasons:

1. **Clarity**: Establishing project parameters provides clarity on what will and won't be included in the migration process, ensuring everyone involved in it works towards similar objectives.
2. **Effective collaboration**: Defined project scope eliminates unnecessary duplication of effort.
3. **Effective resource management**: A clearly defined project scope helps allocate resources efficiently, such as personnel, equipment and funding. This ensures that resources are utilized efficiently and effectively, minimizing the risk of cost overruns or delays.
4. **Facilitates Communication**: Establishing project scope helps communicate between all parties involved, such as executives, project managers, IT personnel and end-users. It guarantees everyone has a shared understanding of the objectives and requirements for the project.
5. **Facilitates Risk Management**: By clearly outlining a project scope, potential risks and challenges can be identified and addressed early on in the process. Doing this minimizes their impact and keeps your project on track.

Key Elements of Defining Project Scope

Establishing project scope ensures key elements such as budget allocation, timelines, resources needed, etc. can all be defined more precisely so you know exactly what you're getting into before any work begins on it.

1. **Goals and Objectives**: Defining the goals and objectives for the project is paramount in creating its scope. This includes specifying specific outcomes and deliverables expected from the migration process.
2. **Deliverables**: Define the deliverables for the project so everyone involved in it understands their responsibilities. This includes identifying data to be migrated, systems involved in the process and any additional documentation needed.
3. **Timelines**: Establishing timelines for a project helps guarantee it remains on track and is completed within its estimated timeline. This includes defining key milestones and deadlines for the job at hand.
4. **Budget**: Determining a project budget is essential for efficiently allocating resources. This includes determining personnel, equipment and funding needed for migration.

Example

To demonstrate the significance of project scope, let's consider a real-life use case:

A large financial services organization is looking to migrate their customer data onto a new system. The scope would include elements such as:

1. **Goals and Objectives**: The objective of the project is to migrate all customer data onto the new system without any loss or disruption in service.

2. **Deliverables**: Included in this deliverable list are all customer records, such as account info, transaction history, and personal details.
3. **Timelines**: The project is expected to take six months to complete, with key milestones and deadlines identified along the way.
4. **Budget**: A total budget of $2 million has been set for this venture - this includes personnel costs, equipment requirements, as well as any extra costs related to migration activities.

By clearly outlining a project scope, financial services organizations can guarantee that the migration process is efficient, successful and meets desired outcomes. It provides a framework for decision-making, efficient resource allocation and project tracking to keep things on track.

3.2 ESTABLISHING PROJECT OBJECTIVES

Establishing project objectives is an integral component of any successful data migration project. They act as a guideline for the team during the migration, offering direction and clarity throughout. In this section, we'll examine why setting project objectives is so important and describe key elements involved in this step-by-step procedure; plus, we'll provide an actual use case to illustrate these concepts.

What Are Project Objectives?

Project objectives are the specific goals and outcomes the data migration project strives to attain. They give direction for the migration process and guarantee all parties involved are working towards same targets.

Why Establishing Project Objectives Matter?

Establishing project objectives is critical in order to successfully complete any data migration task.

Establishing project objectives is essential for several reasons:

1. **Clarity**: Determining project objectives provides clarity on the specific goals and outcomes of the migration process, ensuring all parties involved in it are on-board and working towards similar outcomes. This ensures everyone stays motivated throughout the project.
2. **Measures Success**: By setting project objectives, the success of the migration process can be evaluated against specific goals and outcomes. This helps guarantee that the project meets its desired outcomes and proves to be a success.
3. **Facilitates Decision-Making**: Establishing project objectives helps to streamline decision-making throughout the migration process. It guarantees that all decisions are in line with the overall goals and objectives of the venture.
4. **Encourages Motivation and Engagement**: Clear project objectives provide motivation and engagement for the project team. They guarantee everyone is working towards a common goal and can see progress being made towards it.

Key Elements of Establishing Project Objectives

Establishing project objectives involves:

1. **Define the Scope**: Determining the scope of a project is essential to guarantee it meets its goals and is in sync with those of the organization as a whole. 3. Evaluate Results: Assessing project outcomes is another essential element when setting project objectives.

2. **Determining Stakeholders:** Determining who are the stakeholders in a migration project is essential for ensuring everyone works towards similar objectives.
3. **Define Specific Goals and Outcomes**: It is essential to set specific objectives and outcomes so all parties involved in the project understand their responsibilities, as well as what outcomes are desired.
4. **Formulating a Plan of Action**: It is essential to create an action plan in order to guarantee that objectives are achieved within the expected timeline and budget.

Example

To demonstrate the significance of project objectives, let's take a look at an actual use case: A manufacturing company wants to migrate their production data onto a new system. The project objectives would encompass elements such as:

1. **Delineating the Scope**: The objective of the project is to migrate all production data onto a new system without any data loss or disruption in service.
2. **Identifying Stakeholders**: Key participants in this project include production personnel, IT department personnel and senior management.
3. **Determining Specific Goals and Outcomes**: The project has set specific objectives to increase production efficiency by 20%, reduce downtime by 10%, and boost data accuracy by 15%.
4. **Formulating a Plan of Action**: Create an action plan that includes key milestones and deadlines, resource allocation, and risk management strategies.

Establishing project objectives for manufacturing companies helps guarantee that the migration process is successful and delivers on its promises. It provides direction for the team, facilitates decision making, and keeps everything on track.

3.3 IDENTIFYING STAKEHOLDERS AND THEIR ROLES

Identification of stakeholders and their roles is a fundamental step for any successful data migration project. Stakeholders refer to individuals or groups with an interest or influence over the migration process, who can affect or be affected by it. In this section, we'll look at why identifying stakeholders is so important, their roles in it, as well as some key elements involved in this step-by-step project.

Importance of Stakeholders Identification and Their Roles

It is critical to identify stakeholders and understand their responsibilities.

1. **Project Success**: Accurately identifying stakeholders and their roles is essential for several reasons. Assures everyone affected by the project is included in the migration process.

Doing this helps guarantee everyone's needs and expectations are fulfilled during this critical stage.

2. **Facilitates Communication**: Determining stakeholders and their roles helps facilitate communication between the project team and those affected by it. Doing this ensures everyone stays up to date on project developments, with plenty of opportunity for feedback from everyone involved.
3. **Facilitates expectations management**: Determining stakeholders and their roles helps to manage expectations by ensuring everyone understands what can be expected from the project and their responsibilities. This ensures everyone has a shared understanding of what can be expected from it and their responsibilities within it.
4. **Fosters Buy-in and Support**: Determining stakeholders and their roles is essential for securing buy-in and support for a project. By understanding each stakeholder's role, stakeholders understand their contribution to the project, as well as become invested in its success.

Key Elements of Identifying Stakeholders and Their Roles

Identification of stakeholders involves understanding who they are, what their responsibilities are, and why they're involved. It provides assurance that everyone understands their significance to the project at hand.

Identifying stakeholders requires consideration of several elements, such as:

1. **Determining Internal and External Stakeholders**: Internal stakeholders refer to those within an organization who are directly affected by the project, such as project team members, department heads, and senior management. Conversely, external stakeholders refer to individuals or groups outside the organization that may

have an impact on the project - like customers, suppliers, and regulatory bodies.

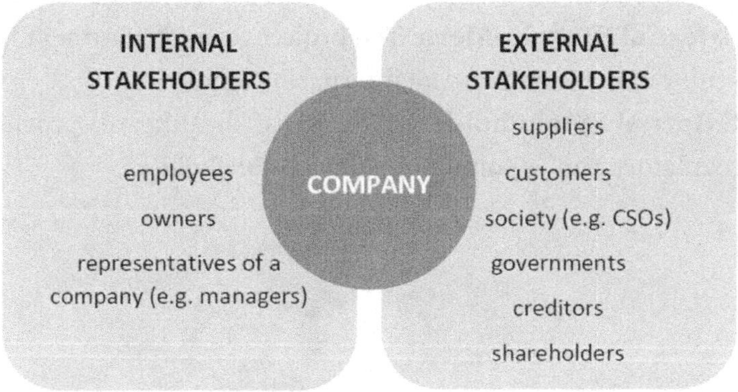

2. **Delineating Stakeholder Roles and Responsibilities**: Establishing stakeholder roles and responsibilities helps guarantee everyone understands their expectations for the migration process, such as project sponsor, manager, subject matter experts and end-users. With this definition in place, all parties involved will be clear about what needs to be done and when.
3. **Determining Stakeholder Needs and Expectations**: Determining stakeholder needs and expectations is essential to guaranteeing the project meets everyone's requirements. This involves collecting feedback from stakeholders, such as which data needs migrating, when it should occur, and how it should be verified.
4. **Crafting a stakeholder engagement plan**: Crafting an effective stakeholder engagement plan helps guarantee that stakeholders are involved throughout the migration process. This involves selecting communication channels, frequency and content.

Example

To demonstrate the significance of identifying stakeholders and their roles, let us look at a real-life use case: A healthcare organization is looking to migrate patient data from an outdated system to a new one. The key stakeholders involved in the project include:

1. **Internal Stakeholders**: The project team, department heads and senior management are all internal stakeholders.
2. **External Stakeholders**: Patients, healthcare providers and regulatory bodies comprise external stakeholders.

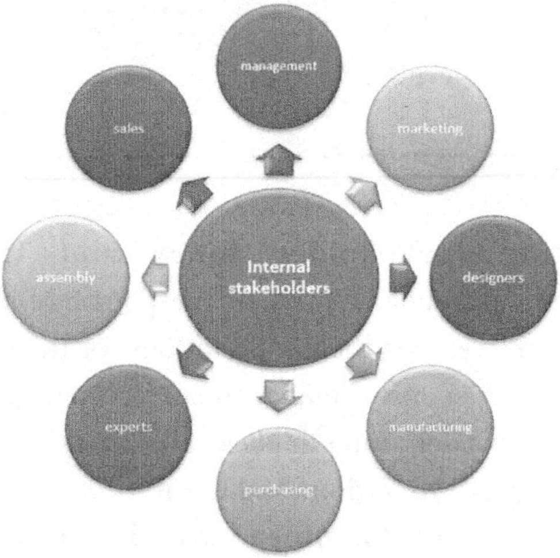

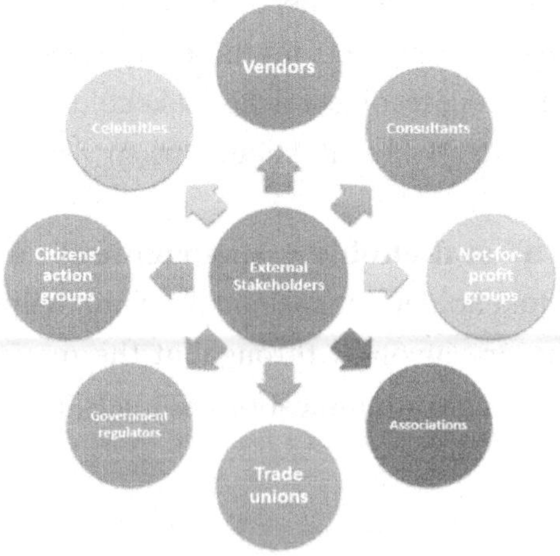

The roles and responsibilities of each stakeholder would then be as follows:

1. **Project Team**: Responsible for planning and executing the migration process.
2. **Department Heads**: Guaranteeing accurate data migration in their department on a timely basis.
3. **Senior Management Team**: Providing resources and support for the project.
4. **Patients**: Provide feedback on the migration process and guarantee their data is migrated accurately.
5. **Healthcare Providers**: Ensure patient data is accurate and updated in a timely manner.
6. **Regulators**: Guaranteeing that the data migration process abides by all regulations.

Determining the needs and expectations of each stakeholder requires collecting feedback on data accuracy, migration timeline, and communication preferences. A stakeholder engagement plan would then be created to guarantee that each participant is kept informed and involved throughout the migration process.

3.4 CHALLENGES IN ESTABLISHING PROJECT SCOPE AND OBJECTIVES

Establishing project scope and objectives are essential steps in any data migration project. They help guarantee everyone involved in the project understands what's expected of them and what should be achieved at the end. But creating such objectives can be a difficult task; there are several challenges that may arise during this stage. In this section, we'll look at some of these key obstacles and discuss strategies for overcoming them.

Lack of Clarity

One of the primary obstacles to setting project scope and objectives is a lack of clarity. This can arise due to inadequate communication between team members and stakeholders, or an inability to comprehend all objectives set out for the project. Without clarity, defining scope accurately and setting achievable goals becomes nearly impossible.

To successfully tackle this challenge, it is essential to communicate clearly with all project stakeholders. This includes outlining the project goals and objectives so everyone comprehends them. It may also be beneficial to create a project scope statement which outlines its scope, objectives, and deliverables; this document can serve as an accessible reference point for all involved throughout the duration of the project.

Limited Resources

Another hurdle in defining project scope and objectives is limited resources. This could include time, budget constraints, or personnel issues. When resources are scarce, it becomes challenging to define the project scope clearly and set achievable objectives that are within budget.

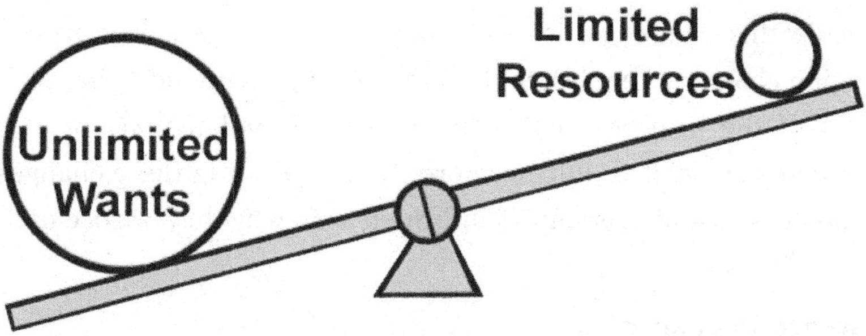

To successfully tackle this challenge, it is essential to prioritize the project's goals and objectives based on available resources. This may mean breaking the project into smaller phases or prioritizing certain objectives over others. Furthermore, communicating resource limits to stakeholders ensure everyone understands what can be accomplished within those constraints.

Changing Requirements

Another challenge when setting project scope and objectives is the potential for requirements to evolve over time. As projects move forward, it is common for existing requirements to shift or new ones may emerge, making it difficult to accurately define the scope of a project and set achievable goals.

To tackle this challenge, it's essential to have a system in place for managing changing requirements. This may involve conducting regular reviews of the project's scope and objectives and making necessary adjustments as needed. Furthermore, communicating these changes to stakeholders ensure everyone is informed as to why they are occurring.

Resistance to Change

Resisting change can be a significant obstacle when creating project scope and objectives. When stakeholders are resistant, it can be difficult to define the project's goals, set achievable milestones, or overcome fear of the unknown. Resistance may stem from various sources such as lack of understanding about the project objectives or concerns about job security.

To successfully tackle this challenge, it is essential to involve stakeholders in the project planning process and guarantee everyone understands its goals and objectives. It may also be beneficial to provide training and support to stakeholders so they can adjust to changes caused by the project. Moreover, regular communication with stakeholders is necessary in order to address any worries or questions they may have.

Establishing project scope and objectives is a fundamental step in any data migration project, yet it's not without its challenges. By communicating clearly with stakeholders, prioritizing goals based on available resources, managing changing requirements, and dealing with resistance to change effectively, these obstacles can be overcome and the project will be successful.

3.5 BEST PRACTICES FOR DEFINING PROJECT SCOPE AND OBJECTIVES

Determining project scope and objectives are essential elements of any data migration project. They lay the framework for success, guaranteeing everyone involved understands their responsibilities and what results to expect at the end. However, this step can be tricky; there are numerous elements to take into account and any missteps now could have major repercussions later on in the process. In this section we'll look at some best practices for crafting project scopes and objectives.

Involve All Relevant Stakeholders

One of the most essential practices when defining project scope and objectives is to involve all relevant stakeholders. This includes anyone who will be affected by or invested in its success. By including stakeholders early on in the process, you can guarantee everyone's needs are taken into account when creating project goals and objectives.

When engaging stakeholders, it is essential that everyone comprehends the purpose of the project and their responsibilities. This can be accomplished through regular communication as well as providing detailed documentation outlining all goals and objectives associated with it.

Establish Clear Boundaries

Determining project scope necessitates setting boundaries around what is included and not included. This can be a difficult task due to competing priorities or differing expectations from stakeholders, but setting clear parameters helps guarantee that resources are used efficiently and on schedule.

Establishing clear boundaries is essential to define the project's objectives and deliverables. This can be accomplished through the creation of a project scope statement that outlines its purpose, goals, objectives, and deliverables. The scope statement should then be reviewed and approved by all relevant stakeholders, so everyone is on the same page.

Establish Measurable Objectives

Setting measurable objectives is the key to success in any data migration project. Measurable objectives allow progress to be tracked and the project remains on track. They should be specific, achievable, relevant, and time-bound.

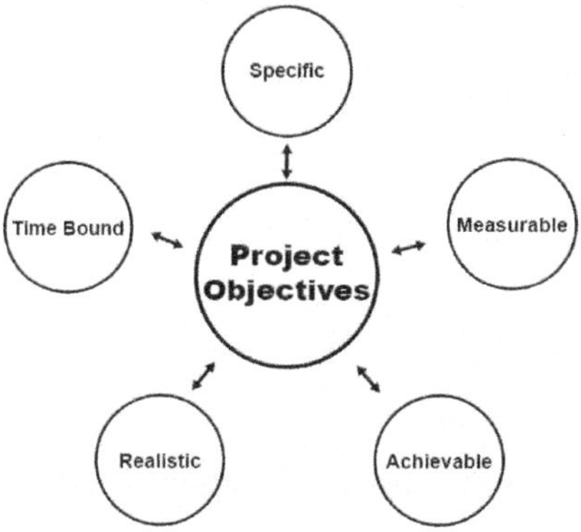

Determining measurable objectives requires specifying specific outcomes that the project should accomplish. These should be connected to its goals and relevant for stakeholders. Furthermore, creating a timeline for achieving these targets is essential.

Regularly Review and Adjust Project Scope and Objectives

Defining project scope and objectives isn't a one-time project. As the project moves forward, it may become necessary to revise these parameters due to new information or shifting circumstances. Regularly reviewing and amending project scope and objectives ensures the project stays on track while making efficient use of resources.

Establishing a process for regularly reviewing and updating project scope and objectives is essential. This may involve having regular meetings with stakeholders to discuss progress as well as any modifications made to the scope or objectives. Furthermore, conducting periodic reviews of progress will guarantee that resources are being utilized efficiently.

Determining project scope and objectives are essential elements of any data migration project. While this may take some time and effort to do, doing so helps guarantee that the project stays on track, meets stakeholder expectations, and meets its goals.

3.6 SUMMARY

- Determining project scope and objectives is a fundamental step in data migration projects.

- Accurately understanding these parameters requires recognizing their boundaries as well as setting clear objectives for the project at hand.

- Defining project scope involves outlining its parameters and limitations, such as goals, deliverables, timeline, budget and resources.

- Establishing project objectives involves outlining measurable yet achievable outcomes that the project aspires to reach.

- Determining stakeholders and their roles is essential when creating project scope and objectives.

- Issues encountered while creating these documents include lack of clarity, inadequate communication, ineffective stakeholder involvement, and resistance to change.

- For effectively defining project scope and objectives, best practices include involving stakeholders, outlining goals and objectives clearly using SMART criteria, taking into account potential risks or constraints, and regularly revising these documents.

3.7 TEST YOURR KNOWLEDGE

I. What is the purpose of establishing project scope and objectives in data migration?

a) To identify potential risks and constraints
b) To define project goals and deliverables
c) To select data migration tools and techniques
d) To create a project schedule and budget

II. What is the project scope?

a) The measurable and achievable goals of the project
b) The resources needed to complete the project
c) The timeline for completing the project
d) The parameters and limitations of the project

III. Why is it important to define project objectives?

a) To create a detailed project schedule
b) To identify potential risks and constraints
c) To ensure stakeholder involvement
d) To define the goals the project aims to achieve

IV. Who are stakeholders in data migration projects?

a) Only project managers and IT staff
b) Only end-users of the data
c) Anyone who has an interest in the project or is affected by it
d) Only executive management

V. What is the role of stakeholders in establishing project scope and objectives?

a) To create the project schedule and budget
b) To identify potential risks and constraints
c) To provide feedback on project goals and objectives
d) To select data migration tools and techniques

VI. What are some common challenges in establishing project scope and objectives?

a) Lack of clarity, poor communication, lack of stakeholder involvement, and resistance to change
b) Lack of budget, inadequate resources, and outdated technology
c) Inability to meet project deadlines, poor quality of work, and lack of project management expertise
d) Poor data quality, lack of data governance, and data security concerns

VII. What are SMART criteria in defining project objectives?

a) Specific, measurable, attainable, relevant, and time-bound
b) Simple, meaningful, actionable, reliable, and timely
c) Strategic, measurable, accurate, reliable, and time-bound
d) Structured, manageable, achievable, realistic, and time-sensitive

VIII. Why is it important to consider potential risks and constraints in defining project scope and objectives?

a) To ensure project deadlines are met
b) To identify opportunities for cost savings
c) To mitigate potential issues that could impact the project's success
d) To streamline the data migration process

IX. What is the benefit of regularly reviewing and revising project scope and objectives?

a) To increase project costs
b) To ensure project deadlines are met
c) To address changes in project requirements and priorities

d) To select data migration tools and techniques

X. What is the role of project managers in establishing project scope and objectives?

a) To make all decisions related to project scope and objectives
b) To work with stakeholders to identify project goals and objectives
c) To select data migration tools and techniques
d) To provide technical support during the data migration process

3.8 ANSWERS

I. Answer: b) To define project goals and deliverables
II. Answer: d) The parameters and limitations of the project
III. Answer: d) To define the goals the project aims to achieve
IV. Answer: c) Anyone who has an interest in the project or is affected by it
V. Answer: c) To provide feedback on project goals and objectives
VI. Answer: a) Lack of clarity, poor communication, lack of stakeholder involvement, and resistance to change
VII. Answer: a) Specific, measurable, attainable, relevant, and time-bound
VIII. Answer: c) To mitigate potential issues that could impact the project's success
IX. Answer: c) To address changes in project requirements and priorities
X. Answer: b) To work with stakeholders to identify project goals and objectives

04

4 ASSESSING DATA QUALITY AND READINESS

Assessing data quality and readiness is a critical step that should not be overlooked during the migration process. This chapter introduces the concept of data quality and readiness, its significance, as well as techniques used to assess its status. In addition, we will explore the metrics used to assess data quality and readiness, their defining criteria, as well as any challenges that may arise during this process. Finally, we will offer best practices that can assist you in conducting a successful data quality and readiness assessment. By the end of this chapter, you will have gained a greater appreciation of how crucial data quality and readiness assessment is during data migration.

4.1 UNDERSTANDING DATA QUALITY & READINESS

Data migration success is heavily dependent upon the quality and readiness of the data being migrated. That is why assessment of data quality and readiness plays such a significant role in any successful migration project.

Data quality refers to the accuracy, completeness, and dependability of data. Inaccurate, incomplete, or unreliable information can lead to errors, inconsistencies, and a loss of confidence in that information. Therefore, assessing data quality prior to migrating it into a new system is imperative for maintaining confidence in its accuracy and dependability.

Data readiness refers to the ability of data to meet the specifications of a new system. It involves verifying that it's in the correct format, contains all essential information, and free from errors or inconsistencies. Assessing data readiness is essential for guaranteeing that migrated information is usable and adds value for the organization.

Before we can understand data quality and readiness, it's essential to first define the criteria. These may include accuracy, completeness, consistency, timeliness, relevance, validity; while data readiness criteria could include format or structure of the data as well as completeness, consistency and conformity. By defining these parameters clearly and objectively, organizations can assess whether their data meets necessary standards for both quality assurance and readiness assurance.

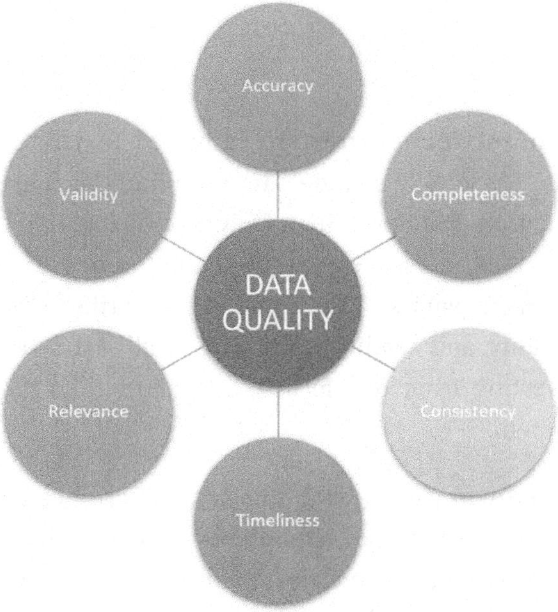

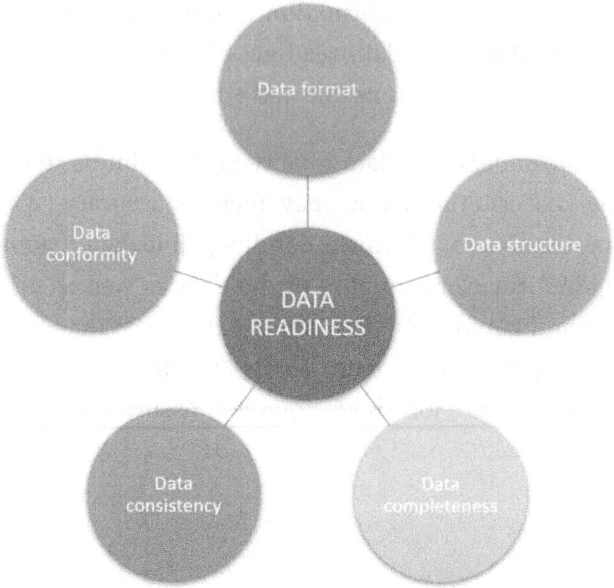

Data assessment techniques are employed to assess data quality and readiness. These may include data profiling, cleansing, validation, and transformation. Data profiling involves analyzing the data to detect patterns, anomalies, and issues with its quality; while cleansing corrects errors or inconsistencies; data validation verifies that it meets predefined quality and readiness criteria; finally data transformation involves converting from one format into another so it's compatible with a new system.

Metrics are used to assess data quality and readiness. Data quality metrics may include accuracy, completeness, consistency, timeliness; while readiness assessment metrics might include data completeness, conformity, consistency and structure. Through metrics organizations can identify areas for improvement in their data handling practices as well as monitor progress towards achieving necessary data quality and readiness goals.

Data quality and readiness assessment remain vitally important, yet organizations may face several obstacles. These could include data complexity, lack of standardization across sources, and the requirement for expertise in data assessment techniques.

To address these challenges, organizations should implement best practices for data quality and readiness assessment. These could include creating criteria for assessment, using automated tools, including stakeholders in the process, as well as regularly monitoring and improving results.

4.2 IMPORTANCE OF DATA QUALITY AND READINESS ASSESSMENT

Data quality and readiness are paramount in data migration projects. Accurate, incomplete, or inconsistent data can cause delays, errors, and even financial loss - which is why conducting an extensive assessment of data quality and readiness before starting any data migration project is so important.

Data quality and readiness assessment are essential components of data migration projects, helping to reduce risks and guarantee data consistency. Here are some of the primary reasons why conducting such an assessment is necessary:

Mitigating Risks:

Data migration projects can be intricate and subject to a variety of potential hazards. One major risk is poor data quality, which could result in incorrect data migration, system malfunctioning, data inconsistencies and incorrect decision-making. By conducting an exhaustive data quality assessment, businesses can reduce these hazards while ensuring their data is migrated efficiently and accurately.

Ensuring Data Consistency:

Data inconsistency can lead to data migration errors, loss, and incorrect processing. It makes managing data more challenging which in turn leads to other issues. A data quality assessment can identify and address inconsistencies, so your data is accurate and consistent across all systems.

Enhancing Data Governance:

Effective data governance is essential for businesses to reach their objectives. Data quality and readiness assessments offer insights into data governance issues such as data ownership, policies, and security that can help businesses identify areas for improvement. By optimizing data governance practices, businesses can maximize data usage while reducing the likelihood of regulatory compliance violations.

Improving Data Quality:

Assessing data quality allows us to detect errors, inconsistencies and inaccuracies that can be addressed before the data migration process starts. Resolving these issues prior to migration helps guarantee that your data is accurate, complete and consistent.

Reducing Costs:

Data quality and readiness assessment can assist businesses in recognizing and fixing issues before the data migration project commences, decreasing the likelihood of delays, errors, or rework. This results in cost savings for businesses by making the migration process more efficient and cost-effective.

4.3 DATA ASSESSMENT TECHNIQUES

Successful data migration relies heavily on the quality and accuracy of the data being migrated. Therefore, it's essential to use effective data assessment techniques in order to guarantee that your information is clean, complete, and consistent before migration begins. In this section we'll review various techniques available for assessing data quality and readiness before migration begins.

Data Profiling:

Data profiling is a technique for analyzing data to discover its characteristics and properties. This analysis involves looking at the structure, content, and relationships within the dataset to detect any inconsistencies or errors. With data profiling you can detect missing data, duplicate records, invalid entries as well as patterns and relationships within it - invaluable assets in data mapping projects.

Example:

In 2018, a major retail company decided to migrate their customer data into a new CRM system. Prior to the migration, they wanted to ensure that all information was accurate and clean for an effortless transition into the new system.

The initial step was to perform data profiling on existing customer information. The team used various assessment techniques such as statistical analysis and data visualization tools to gain insights into the quality of this data, looking at things like missing values, duplicate records, and inconsistent data formats to identify areas for concern.

One of the major issues they discovered was that many customer addresses were incomplete or inaccurate, which could lead to shipping and delivery problems in the future. To address this problem, they devised a data cleansing plan which included standardizing address formats and validating addresses against an established source.

Another issue they identified was that some customer records were missing key information such as email addresses or phone numbers. To remedy this, the team devised a plan to collect these details from customers through an effective email campaign and data entry efforts.

Data Sampling:

Data sampling is a technique that involves selecting an accurate subset of data from a larger set. This approach helps assess the quality and readiness of information by analyzing sample data. Data sampling can be especially beneficial when dealing with large datasets since it allows faster assessment of the information; however, it's essential that the sample data be representative of all samples within it.

Example:

A major financial institution used data sampling to assess its data prior to embarking on a major migration project. They planned to migrate legacy information onto a new platform and wanted to guarantee that the data was clean and accurate before beginning the transition process.

To do this, the institution selected a representative sample of its data and utilized data profiling techniques. This process involved selecting random samples from across various sources within the institution's data warehouse and then analyzing it to detect any quality issues or discrepancies. Specifically, completeness, accuracy, consistency, and conformity to data standards were assessed for all records.

Data sampling revealed several data quality issues that needed to be addressed before migration could begin. For instance, analysis revealed inconsistent customer data storage across various systems and incomplete or inaccurate fields. With this insight, the institution was able to devise a data cleansing strategy which addressed these concerns and permit continued progress with the migration project. This case study highlights the value of using data sampling as part of any data assessment process.

Data Cleansing:

Data cleansing is a technique for detecting and correcting errors and inconsistencies in data. This includes eliminating duplicate records, correcting spelling mistakes, and standardizing data formats. Data cleansing helps improve the quality of your information while decreasing the risk of errors during migration.

Data Reconciliation:

Data reconciliation is a technique for comparing data across different systems to detect any discrepancies. It involves comparing the source system's data with that in the target system to ensure they match up accurately. Data reconciliation can assist in detecting missing or incomplete records during migration processes.

Data Mapping:

Data mapping is a technique that involves creating a mapping between the data in one system and its target system. This involves identifying fields within the source system and mapping them to their corresponding fields in the target system. Data mapping helps guarantee accurate migration of information from one system to another.

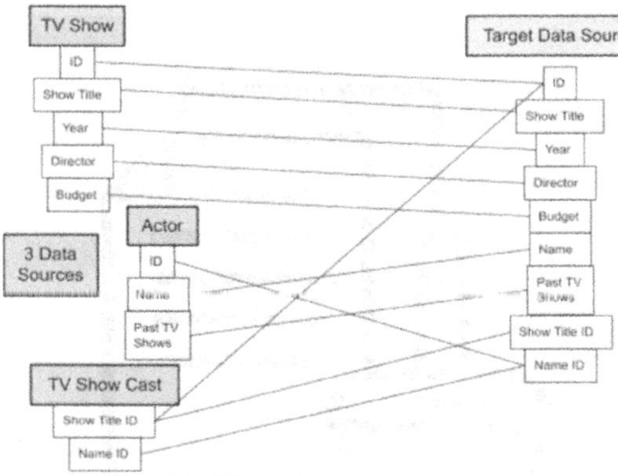

Data Validation:

Data validation is a technique that verifies the accuracy and completeness of data. This involves running tests on it to detect any errors or discrepancies. Data validation helps guarantee that your information is correct before migration so you can move forward confidently with your projects.

4.4 DATA QUALITY METRICS

Data quality is of the utmost importance in data migration projects. Poor data quality can result in inaccurate or incomplete results, leading to incorrect business decisions being made. To guarantee consistent quality throughout a data migration project, robust metrics for data quality must be utilized.

This section will provide a comprehensive overview of data quality metrics, including their definition, types and significance in data migration projects. Furthermore, we'll examine how data quality metrics can be employed to enhance data quality during a data migration project.

Definition of Data Quality Metrics:

Data quality metrics are a set of parameters used to assess the quality of data. They can be used to assess accuracy, completeness, consistency and timeliness. Data quality metrics offer a quantitative measure of data's integrity that helps evaluate it and identify areas for improvement.

Types of Data Quality Metrics:

Data quality metrics can be used to assess data. Some of the most widely employed include:

1. **Completeness**: This metric measures the percentage of data present in a dataset. It helps identify missing records and guarantees all essential information is available.

2. **Accuracy**: This metric assesses the correctness of data. It helps detect errors and guarantee accuracy.
3. **Consistency**: Measures uniformity among data points. It helps detect inconsistent patterns, ensuring all records are consistent.
4. **Timeliness**: This metric assesses the currency of data. It's used to guarantee that information provided is both up-to-date and relevant.

Importance of Data Quality Metrics:

Data quality metrics are vital for maintaining data quality during a data migration project. They offer an objective measure of data quality that can be used to assess existing conditions and pinpoint areas for improvement. Data quality metrics enable us:

1. **Increase Data Accuracy**: Utilizing data quality metrics to detect errors in data can enhance its accuracy, leading to more precise business decisions being made.
2. **Guarantee Data Completeness**: Data quality metrics can be utilized to detect missing data, which then gets added back into the dataset for completeness assurance. This ensures all necessary information is available.
3. **Enhance Data Consistency**: Data quality metrics can be utilized to detect inconsistencies in data, which then allow corrections to be made so the information remains consistent.
4. **Ensuring Data Timeliness**: Data quality metrics can help guarantee the data is up-to-date and relevant, which is essential for making accurate business decisions.

Challenges with Utilizing Data Quality Metrics:

Data quality metrics present several challenges during a data migration project. These obstacles include:

1. **Data Volume**: Excessive amounts of data can make it challenging to analyze and assess data quality metrics accurately. As evidenced by the below charts, data creation has been growing exponentially over the last several years.

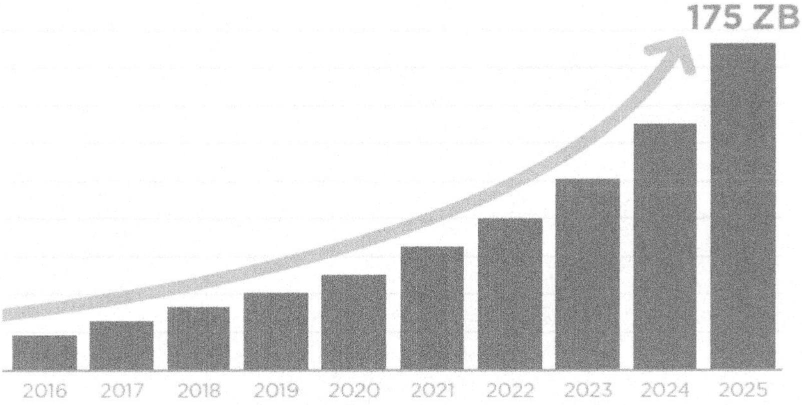

2. **Data Complexity**: Data complexity refers to the degree to which data is difficult to comprehend or analyze due to its volume, variety or velocity. Big data, which includes both structured and unstructured information from various sources, serves as a prime example of data complexity. As this mass of information grows larger and more diverse, analyzing it becomes increasingly challenging; traditional quality metrics may not be adequate in recognizing issues with accuracy and completeness in the collected datasets.
   One of the primary difficulties when using data quality metrics with complex data is a need for a more holistic approach to assessment. Traditional metrics, such as completeness and accuracy, may not provide an accurate representation of data quality in such environments. Therefore, additional criteria like consistency, timeliness and relevance must be developed and

applied; this necessitates an in-depth comprehension of both the data itself and its business context in which it's used.

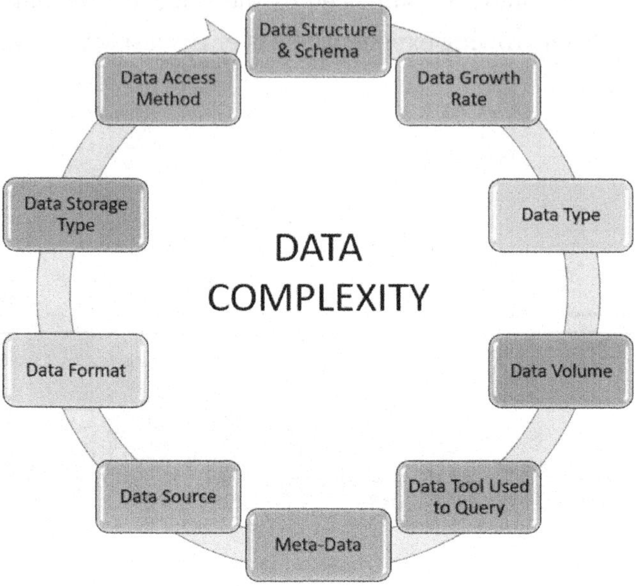

Another challenge lies in the need for more sophisticated tools and technologies to support data quality assessment. These include data profiling, cleansing, and governance tools. These instruments can help identify issues with data quality and offer recommendations for improvement.

Let us consider a real-world example to illustrate the difficulties associated with data complexity and quality metrics: A major retail chain wanted to analyze sales information in order to gain insight into customer behavior and preferences. The company had collected data from various sources, such as point-of-sale systems, customer loyalty programs and online sales. Unfortunately, the information was inconsistent, incomplete and filled with mistakes. Traditional data quality metrics, such as completeness and accuracy, weren't sufficient to address the problems with the data. To address them effectively, the company needed a more comprehensive approach which took into account data consistency, timeliness, relevance as well as implementing tools for profiling and cleansing to improve its quality.

3. **Data Source Variability**: Data can come from various systems and databases, making it difficult to guarantee consistency across all sources.
4. **Data Governance**: Data governance policies and procedures can have an impact on the implementation of data quality metrics, so they need to be taken into account during this step-by-step assessment.

4.5 READINESS ASSESSMENT METRICS

When a company or organization plans to migrate its data, it's essential that the data is prepared for the transfer process. Assessing the readiness of your data and recognizing any potential issues that could occur during migration are two critical tasks that organizations must ensure they do. Utilizing readiness assessment metrics can help organizations guarantee their data is ready for migration and any issues are identified and addressed promptly.

What Are Readiness Assessment Metrics?

Readiness assessment metrics refer to a set of tools and techniques used to evaluate data's readiness for migration. They measure quality, completeness, and accuracy in the data as well as identify any potential issues during the migration process.

Some common examples of readiness assessment metrics include:

1. **Data Completeness**: This metric measures the percentage of data that is complete and accurate. Incomplete or inaccurate data can cause issues during migration, so it's essential to

guarantee all information is verified as complete and correct prior to migration.

2. **Data Consistency**: This metric assesses the consistency of data across different systems or databases. Inconsistent data can lead to issues during migration, so it's essential that all information is accurate prior to moving.
3. Data Quality: This metric assesses the quality of data, taking into account factors like accuracy, completeness and consistency. Accurate information is key for successful migrations so it's essential that all data meets a certain standard before migration begins.
4. **Data Accessibility**: This metric assesses data accessibility, such as its ability to be accessed and retrieved from different systems or databases. Unaccessible data can cause issues during migration, so it's essential that all pertinent information be made readily accessible before moving.
5. **Data Security**: This metric assesses the security of data, taking into account elements like encryption, access controls and protection. Ensuring data security is paramount for successful migration; thus it must be ensured all information is encrypted before transition.

Importance of Readiness Assessment Metrics:

Utilizing readiness assessment metrics is essential for guaranteeing data is fit for migration. Without them, organizations could face issues such as incomplete or inaccurate data, inconsistent information across different systems, poor data quality, inaccessible data and secure information.

Utilizing readiness assessment metrics, organizations can detect any potential issues during the migration process and take immediate steps to resolve them. Doing this helps guarantee a smooth and successful migration, while decreasing the risk of data loss or other issues.

One **real-life example** of the importance of readiness assessment metrics is a migration of customer data for a large retail company. Before beginning the transfer, they conducted an assessment to evaluate the quality of their database and identify any potential issues. As you can imagine, this was no small task!

During their assessment, the company discovered that the data was inconsistent and contained duplicates and missing information. By recognizing these problems ahead of time, they could address them and guarantee the data would be suitable for migration - decreasing the risk of data loss or corruption and guaranteeing a successful migration.

In addition to assuring the quality of data, readiness assessment metrics can also help identify potential risks and challenges during migration. For instance, if an assessment reveals complex or difficult-to-migrate data, organizations can take steps to address these problems before beginning migration - this helps guarantee completion on schedule and within budget.

Challenges with Readiness Assessment Metrics:

Organizations may face several difficulties when using readiness assessment metrics. These include:

1. **Absence of Standardization**: Organizations lack a common set of readiness assessment metrics to use, making it difficult to compare and evaluate data across different systems or databases.
2. **Data Complexity**: Data that is complex can make it challenging to assess its readiness for migration, especially in organizations with large volumes of information or spread across multiple systems or databases. This issue becomes especially challenging in such instances.
3. **Data Quality Issues**: Assessing data quality can be a time-consuming and resource-consuming task, particularly if there are existing data quality problems that need to be addressed before migration. Unfortunately, this process often necessitates significant effort and resources in order to be successful.
4. **Limited Resources**: Assessing data readiness can be a time-consuming and laborious process that necessitates extensive resources, especially for organizations with limited financial resources. This presents an obstacle for those organizations.

4.6 DEFINING DATA QUALITY AND READINESS CRITERIA

Data migration is a complex process that necessitates careful planning, execution and monitoring. One of the key aspects of data migration is guaranteeing the data being migrated is of high quality and that the systems or applications receiving it are prepared. Defining data quality and readiness criteria is an integral part of this project; in this section we'll cover why defining such criteria is so important, discuss potential challenges encountered while outlining best practices for creating them.

Importance of Determining Data Quality and Readiness Criteria:

Determining data quality and readiness criteria is paramount for the success of any data migration project. These standards establish a benchmark for evaluating the quality and readiness of data, systems, and applications involved in the transfer. By adhering to these specifications, you ensure that your migrated information meets the demands of the new system and it's ready to accept it.

Establishing data quality and readiness criteria helps organizations identify potential issues early in the migration process. By setting criteria, organizations can detect data quality and readiness issues before beginning migration operations. Doing this enables them to address these matters before becoming major problems that could cause delays or result in data loss.

Data Quality Criteria:

Data quality criteria are used to assess the quality of migrated data. These may include elements like accuracy, completeness, consistency, validity and timeliness. Establishing data quality criteria helps organizations ensure their migrated information meets all necessary requirements for the new system.

Accuracy: Data accuracy is paramount for ensuring the migrated data is reliable and can be used for decision-making. You can evaluate the accuracy of your data by comparing it with a trusted source or using data profiling tools.

Completeness: Data completeness refers to the presence of all required data elements. Determining completeness criteria helps organizations guarantee that all necessary data is included in a migration project.

Consistency: Data consistency refers to the absence of duplicate or conflicting data. Establishing consistency criteria helps organizations guarantee that migrated information is accurate, free from duplicates or conflicts, and free from mistakes.

Validity: Data validity refers to the conformance of data to predefined rules or standards. Establishing validity criteria helps organizations guarantee that migrated information meets the demands of their new system.

Timeliness: Data timeliness refers to the accuracy and up-to-dateness of data. Establishing timeliness criteria helps organizations guarantee that migrated information remains current and pertinent.

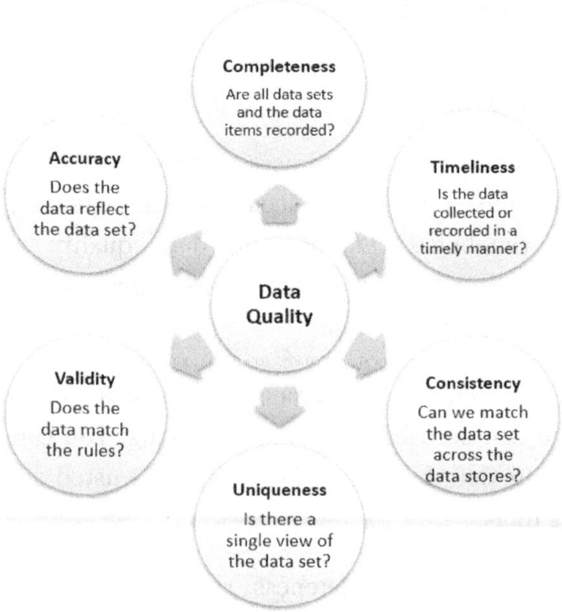

Readiness Criteria:

Readiness criteria are used to assess the readiness of systems and applications involved in migration. These may include factors like hardware/software compatibility, system performance, security, etc. Establishing readiness criteria helps organizations guarantee that their new system is ready to receive migrated data with no issues.

Hardware and Software Compatibility: Successful migration requires that the new system is compatible with the hardware and software being used in the old one. Establishing compatibility criteria helps organizations guarantee that their new system will run seamlessly on top of what was previously in place, whether that means using compatible parts from an original system or creating custom modifications to meet current requirements.

System Performance: The performance of the new system is essential for guaranteeing that migrated data can be processed and used efficiently. Determining performance criteria helps organizations guarantee their new system can handle workloads effectively and perform at required levels.

Security: Data security is paramount for protecting an organization's data from unauthorized access or theft. Establishing security criteria helps organizations guarantee that their new system meets required standards for protection.

4.7 CHALLENGES IN DATA QUALITY AND READINESS ASSESSMENT

Data migration is a complex process that must be overcome in order to guarantee its successful conclusion. One of the biggest hurdles to overcome is assessing data quality and readiness; this involves identifying relevant metrics and criteria, evaluating data against those standards, and making decisions regarding what should be done with any incomplete or inaccurate information.

In this section, we will identify some of the most frequent challenges encountered during data quality and readiness assessment and offer strategies for overcoming them.

1. **Data Accuracy/Impact of Human Error or System Glitches**: When data is incomplete or inaccurate, it can present one of the biggest obstacles when assessing data quality and readiness for migration. This can happen due to human error, system malfunctioning or problems with data sources. When this type of issue arises, making informed decisions about its suitability for migration becomes impossible, leading to delays, errors and other problems that negatively affect the entire migration process.

   To tackle this challenge, it is essential to identify the sources of incomplete or inaccurate data and take steps to rectify them. This may involve improving data entry processes, updating sources of information, or implementing data validation checks that detect errors before they cause issues.

2. **Data Security Concerns**: Another challenge when assessing data quality and readiness is guaranteeing the security of sensitive or confidential data. During assessment, it may be necessary to share this info with third-party vendors or other stakeholders involved in migration processes - which poses security risks since sensitive info could be targeted for theft or hacking.

   To reduce this risk, it is essential to create effective protocols for handling sensitive data during assessment. This may involve encrypting files, restricting access only to those who need them, and monitoring data access and use in order to detect and prevent any unauthorized activity.

3. **Absence of Specific Metrics or Criteria**: One final challenge when assessing data quality and readiness is the lack of specific metrics or criteria for evaluation. Without clearly defined standards for data quality and readiness assessment, it can be difficult to make informed decisions about which data to include in the migration process and which ones to discard.

   To address this challenge, it is essential to create objective and measurable metrics for data quality and readiness. These should be based on industry standards and best practices and tailored specifically for the migration project's needs. Doing this will guarantee all stakeholders are working from similar criteria so they can make informed decisions about the data.

4. **Difficulty in Handling Large Volumes of Data**: Assessing data quality and readiness can be complicated by the difficulty managing large volumes of information. Data migration projects typically involve moving vast amounts of information from one system to another, making it difficult to manage and evaluate this volume accurately.

   To address this challenge, it is essential to create a structured and organized approach to managing data during the assessment process. This could involve breaking up large datasets into smaller subsets for evaluation, using automated tools to assess data quality and readiness, or implementing data governance processes that guarantee consistency and accuracy across all data sets.

5. **Resistance to Change**: Finally, one common obstacle when assessing data quality and readiness is resistance to change. People can be stubborn when it comes to their data; employees may feel bound by tradition or be skeptical of its potential

benefits. Migration of this kind of information requires patience, persistence, and trust - qualities which cannot be guaranteed with traditional methods.

To successfully tackle this challenge, it is essential to involve stakeholders in the assessment process and communicate the advantages of data migration clearly and persuasively. This may involve providing training and assistance to employees as they adjust to new systems and processes, or engaging with stakeholders directly to understand their concerns and address them promptly.

4.8 BEST PRACTICES FOR DATA QUALITY AND READINESS ASSESSMENT

Data quality and readiness assessment is a fundamental aspect of data migration planning and execution. Organizations must guarantee the data they migrate is of high quality and ready for use in the target system, which requires following best practices that can help them detect potential data quality or readiness issues early on and mitigate them before the migration process starts.

In this section, we will outline the best practices for data quality and readiness assessment that organizations can utilize to guarantee a successful data migration.

Development of a Data Quality and Readiness Framework

The initial step to guaranteeing data quality and readiness is creating a framework that sets forth criteria for assessing both. This should specify metrics for assessment, aligned with organizational business goals, as well as include guidelines on data cleansing/transformation procedures as well as data profiling to detect any issues with quality.

Identify Data Sources

Organizations must identify all data sources that will be migrated and assess their quality and readiness. This may involve data profiling, cleansing, and transformation activities to guarantee that the data is suitable for use within the target system.

Perform a Data Inventory

A data inventory is an essential step in data quality and readiness assessment. It involves compiling an exhaustive list of all data to be migrated, including its format, structure, and location. By conducting this inventory, organizations can identify any issues with data quality or readiness that need to be addressed before beginning the migration process.

Define Data Quality and Readiness Metrics

Organizations must create data quality and readiness metrics that align with their business goals. These measures should be used to assess the data's quality before and after migration, be they well-defined, measurable, and related to the organization's core targets.

Develop Data Cleaning and Transformation Procedures

Organizations need to devise data cleansing and transformation procedures that can assist them with addressing data quality and readiness concerns. These should be based on the metrics defined in the framework, with a goal of ensuring that the data is fit for use within the target system.

Testing Data Quality and Readiness

Organizations must test the data quality and readiness prior to migrating it to a target system. This involves running test scenarios that simulate the actual migration process, verifying that all data meets predefined quality and readiness metrics.

Monitor Data Quality and Readiness

Organizations must continue to monitor data quality and readiness even after the migration process has been completed. To do this, organizations need to set up monitoring tools that can alert them of potential data quality or readiness issues. The monitoring process should be well-defined and integrated with an organization's business processes.

Engage Stakeholders

Data quality and readiness assessment is a collaborative effort that involves stakeholders from across the organization. These could include business owners, IT personnel, data stewards, and data analysts. Involving these parties in this process helps organizations identify potential data quality or readiness issues and guarantee that the data is suitable for use within their intended system.

4.9 SUMMARY

- Recognizing the significance of assessing data quality and readiness before migration
- Utilizing techniques such as sampling, profiling, and data mapping
- Measuring key data quality indicators such as completeness, accuracy, consistency, and integrity

- Readiness assessment metrics such as infrastructure, resources and stakeholder engagement

- Determining data quality and readiness criteria based on project requirements

- Challenges associated with data quality and readiness assessment such as lack of resources or limited data access

- Best practices for data quality and readiness assessment involve engaging stakeholders, using automated tools, and conducting regular assessments.

4.10 TEST YOUR KNOWLEDGE

I. What is the main purpose of assessing data quality and readiness in data migration?

a) To waste time and resources
b) To identify potential data issues
c) To reduce efficiency of the migration process
d) None of the above

II. What are some of the data assessment techniques used in data migration?

a) Data profiling
b) Data mapping
c) Data sampling
d) All of the above

III. Which of the following is not a data quality metric?

a) Accuracy
b) Completeness
c) Consistency

d) Speed

IV. Which of the following is a readiness assessment metric?

a) Completeness of data
b) Data migration time
c) Data profiling
d) Data mapping

V. Why is it important to define data quality and readiness criteria?

a) To ensure consistency
b) To establish clear expectations
c) To reduce efficiency
d) None of the above

VI. What are some challenges in data quality and readiness assessment?

a) Insufficient data governance
b) Poor data quality
c) Limited resources
d) All of the above

VII. What is a best practice for data quality assessment?

a) Using multiple data sources
b) Relying on a single data source
c) Ignoring data outliers
d) None of the above

VIII. What is a best practice for data readiness assessment?

a) Communicating clearly with stakeholders
b) Ignoring stakeholder input
c) Not involving stakeholders
d) None of the above

IX. What is the role of data mapping in data quality assessment?

a) To identify potential data issues
b) To ensure data completeness
c) To establish clear expectations
d) None of the above

X. Which of the following is not a challenge in data quality and readiness assessment?

a) Limited resources
b) Insufficient data governance
c) Clear expectations
d) Poor data quality

4.11 ANSWERS

I. Answer: b) To identify potential data issues
II. Answer: d) All of the above
III. Answer: d) Speed
IV. Answer: a) Completeness of data
V. Answer: b) To establish clear expectations
VI. Answer: d) All of the above
VII. Answer: a) Using multiple data sources
VIII. Answer: a) Communicating clearly with stakeholders
IX. Answer: a) To identify potential data issues
X. Answer: c) Clear expectations

05

5 DEVELOPING A MIGRATION STRATEGY

Data migration can be a complex process that presents both technical and logistical obstacles. Therefore, organizations must devise an organized and efficient migration strategy in order to successfully execute this task. In this chapter, we'll identify the essential elements of crafting a migration strategy, such as an understanding of data migration strategy, its significance, phases and types, selecting the correct strategy type, process requirements, risks mitigations identified during migration, timelines established for data mapping techniques and transformation techniques used. By having this foundational knowledge organizations can guarantee successful and efficient data migration that meets their individual needs and specifications.

5.1 UNDERSTANDING DATA MIGRATION STRATEGY

Data migration is a complex undertaking that necessitates an organized strategy for success. Crafting such a plan is essential to effectively and efficiently moving data between systems. It serves as a roadmap that guides the migration, helping minimize risks, minimize downtime, and guaranteeing accurate migration of your sensitive information securely.

Understanding the key components of a data migration strategy is critical for anyone involved in an operation. A thorough comprehension can help you make informed decisions and anticipate potential issues before they arise. In this section, we'll give an extensive explanation of what a data migration strategy is and how it can be utilized to ensure successful migration results.

What is Data Migration Strategy?

A data migration strategy is an outline of how data will be moved from one system or platform to another. It includes processes, procedures and best practices designed for successful and secure data migration. A well-crafted migration plan should take into account all critical aspects such as data mapping, quality assurance and testing in order to guarantee successful transition of your information.

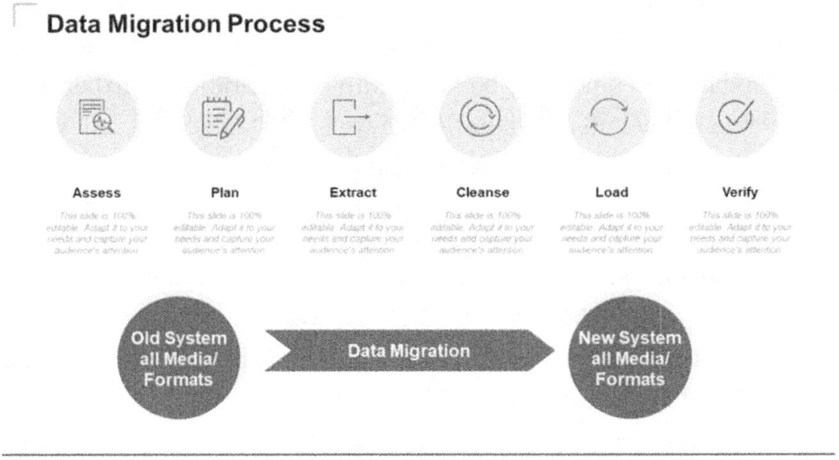

The data migration strategy should also define a timeline, identify key stakeholders, and clearly define each team member's roles and responsibilities. It should be in line with the organization's overall objectives and goals while taking into account any compliance or regulatory requirements.

Example:

A company is moving its data from an outdated legacy system to a modern platform that supports modern technologies and features.

There is an immense amount of information that must be transferred, including customer information, financial data, and product details. To guarantee an effortless migration process without any data loss or corruption during the transition, the IT department needs to devise a data migration strategy.

To create a data migration strategy, the IT department begins by assessing the company's current data architecture and identifying which data needs to be moved. They also consider data dependencies like how it is stored, accessed, and shared across different systems and departments.

Next, they define the scope of the migration, including which data sets will be migrated and in what order. Any dependencies or relationships between data sets must also be taken into consideration. Furthermore, they create a timeline for migration as well as establish key milestones and checkpoints to guarantee it stays on track.

The IT department then determines the optimal migration strategy based on an analysis of their current data architecture and scope of the move. They may opt for a big bang approach, in which all data is transferred simultaneously, or opt for a phased approach where smaller batches of data are migrated over time.

Once a migration strategy is in place, IT departments create data mapping and transformation techniques to guarantee accurate migration of the data. Furthermore, they identify any potential risks or issues that may arise during this process and devise mitigation tactics to address them.

By understanding their data migration strategy, companies can successfully migrate their information without causing significant disruption to operations and guarantee the integrity and accuracy of their data throughout the transition.

5.2 IMPORTANCE OF DEVELOPING A MIGRATION STRATEGY

A migration strategy is a comprehensive plan that outlines the steps involved in moving data from one system to another. It includes selecting appropriate tools, assessing risks and challenges involved, as well as setting an achievable timeline for migration. Here are some reasons why developing such a strategy is so critical:

Minimizes Downtime and Business Disruption

Data migration can be a complex process that may cause system downtime and business disruption if not carefully planned and executed. Crafting an effective migration strategy helps minimize downtime, while guaranteeing that operations don't get disrupted during the transfer. Businesses with an established plan in place are better able to detect potential issues that could cause system downtime ahead of time and take proactive measures to avoid them from occurring.

Gartner estimates that downtime can be costly, with an average cost of $5,600 per minute. Furthermore, 98% of organizations estimate a single hour of downtime to cost over $100,000. Daily interruptions take an average of 238 minutes and require an additional 84 minutes for regular operations to resume; this results in an annual cost estimate for the U.S. of $588 billion. Small businesses affected by severe events like hacking or ransomware face the potential risk of closure within six months; 60% often do. Even minor incidents cause substantial harm in terms of downtime, loss of productivity, and expenses alike - making even minor incidents costly in terms of damages such as lost productivity and expenses incurred from these incidents.

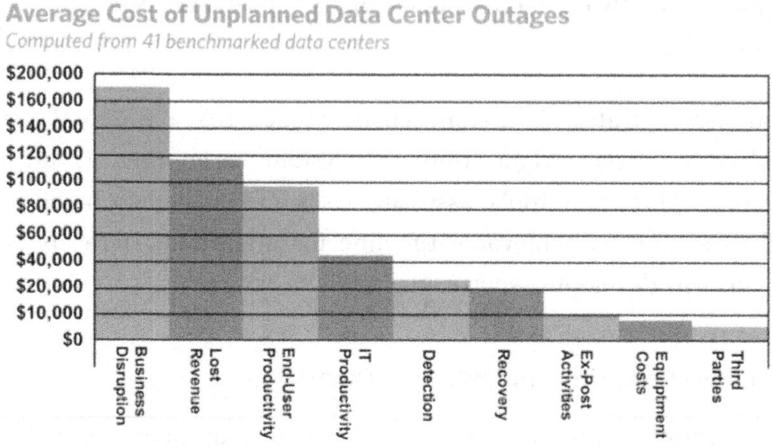

Source: https://calyxit.com/the-real-cost-of-business-interruption/

Reduces Data Loss

Migrating data between systems carries a high risk of data loss. A migration strategy helps minimize this danger by creating an extensive plan that outlines all steps involved in the transfer. It should include data backup and recovery procedures, guaranteeing data is safeguarded in case any unexpected issues arise during transit.

Data Accuracy and Integrity

Migrating data from one system to another poses a risk of data corruption or integrity issues. A migration strategy helps minimize these risks by creating an action plan that guarantees accurate transfer of information between systems. The strategy includes data validation and verification processes, guaranteeing data is free from errors or inconsistencies.

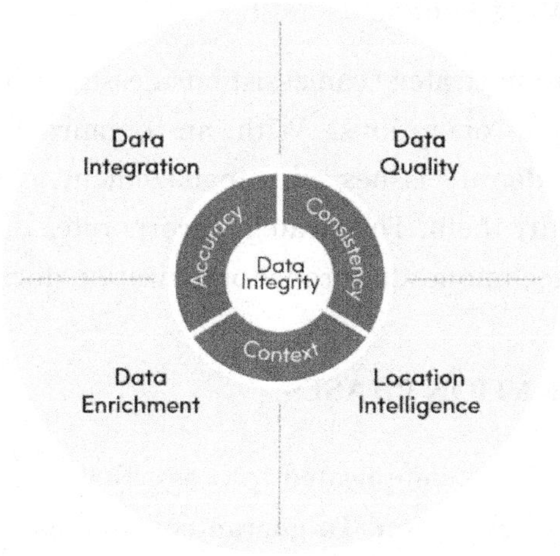

Saves Time and Costs

Crafting a migration strategy can help businesses save time and expenses during the migration process. A migration strategy lays out an organized plan that outlines all steps involved, helping business identify potential issues which could cause delays or additional expenses. With an effective plan in place, companies can take proactive measures to avoid problems and minimize both time and costs related to migration.

Improves Data Management

Crafting a migration strategy can assist businesses in streamlining their data management operations. With an organized plan in place, companies can identify issues with management and take proactive measures to rectify them. The strategy incorporates data mapping and transformation techniques for proper organization during migration.

5.3 DATA MIGRATION PHASES

Data migration is a complicated process that necessitates careful consideration and execution. To guarantee a successful migration, it's essential to comprehend the various phases involved. In this section, we will take an in-depth look at these various data migration stages.

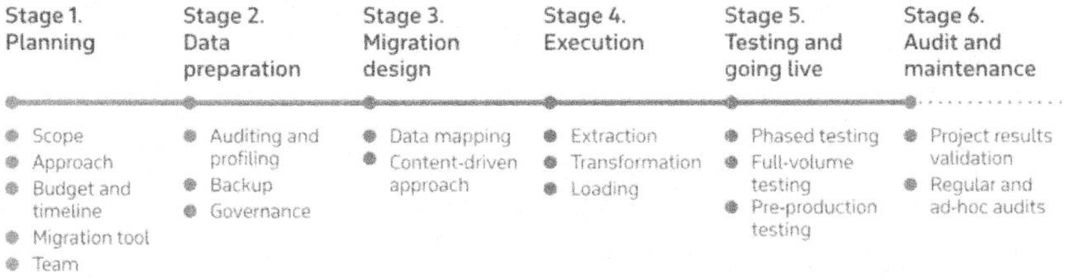

Planning Phase

The planning phase is the initial step in data migration. It involves defining the scope, data to be migrated, and resources needed for completion. After this has been identified, a project plan is created outlining all necessary steps that need to be taken during migration.

Analysis (Preparing) Phase

The analysis phase involves a comprehensive examination of the data to be migrated, helping identify any issues or challenges that could arise during this step. Data is assessed for completeness, accuracy, and consistency to ensure there are no problems during migration. Any quality issues should then be identified and addressed prior to beginning the actual migration procedure.

Design Phase

In the design phase, a data migration strategy is created. This involves selecting suitable tools and techniques, determining how data will be migrated, as well as outlining the migration process. Additionally, data mapping documents are created that describe how transformed and migrated data from one system to another.

Build (Execution) Phase

The build phase is when the actual migration begins. Here, data is extracted from the source system, transformed to meet requirements of the target system, and loaded into it. Additionally, testing of the migration process helps confirm everything works as expected during this stage.

Test Phase

The testing phase includes a rigorous examination of the migration process. This includes both functional and performance checks to guarantee it functions optimally and efficiently. Any issues identified during this evaluation are rectified prior to finalizing the transfer.

Deploy Phase

The deploy phase is the final step of data migration. Here, migrated data is validated and tested to meet requirements for the target system before being made accessible to end-users. Additionally, monitoring is done on this stage to guarantee everything works as expected.

Example:

Let's consider the case of a large e-commerce company that wants to migrate its customer data from its current legacy system to a modern cloud-based one. As the business has expanded rapidly in recent years, its current system cannot handle the growing volume of customer information.

To guarantee a successful migration, companies should adhere to an organized data migration process that includes several stages:

1. **Discovery**: During this phase, the company's IT team will identify all data that needs to be migrated, such as customer information such as names and addresses as well as order history. They'll also assess its quality and identify any potential risks or challenges associated with migration.
2. **Analysis**: Once the data has been identified, the IT team will examine it to verify its accuracy, completeness, and consistency. They'll also identify any duplicate or redundant records and remove them to guarantee the migrated information is as clean as possible.
3. **Design**: In this phase, the IT team will create the target data model and map data between source and destination systems. Doing so guarantees that data is migrated accurately, with the target system configured to handle its new found content.
4. **Development**: Once the design is approved, the IT team can begin creating migration scripts and data transformation rules. This step also involves setting up the target system, configuring data migration tools, and testing out the migration process.

5. **Testing**: During this phase, the IT team will thoroughly test the migration process to guarantee that migrated data is accurate and complete. They also check to see if the target system functions as expected and can handle the migrated information.
6. **Deployment**: Once testing is complete, the IT team will launch the migration process to transfer data from the legacy system to the desired target system. They'll monitor this step closely in order to prevent errors or data loss during transfer.
7. **Post-migration Review**: Once the migration is complete, the IT team will conduct a post-migration review to confirm that all data transferred was accurate and comprehensive. They also perform any necessary data clean-up or optimization tasks so that the target system is optimized for success.

By adhering to a well-defined data migration process that includes these phases, an e-commerce business can guarantee successful migration with minimal disruption to their operations and establishes a strong foundation for future growth.

5.4 MIGRATION STRATEGY TYPES

As data volumes continue to expand, organizations are seeking ways to optimize their data management and storage systems. Data migration is one such solution that enables organizations to move data from one location to another, upgrade existing systems, or consolidate multiple sources into a centralized location. But selecting the correct migration strategy for your organization's requirements is paramount; in this section we'll explore different types of migration strategies along with their benefits and drawbacks.

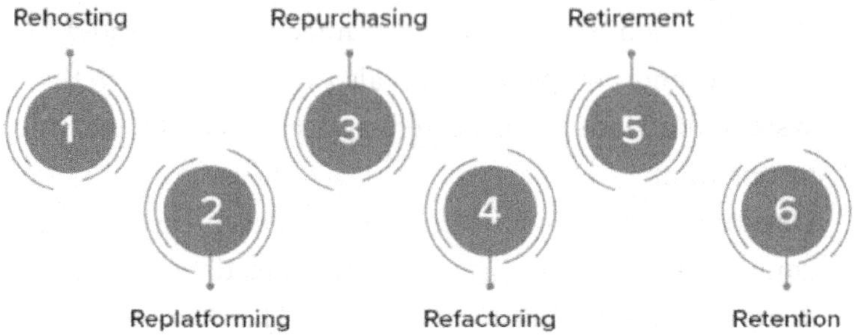

Big Bang Migration

A big bang migration strategy involves migrating all data from one system to another at once. This method works best for smaller organizations or those with limited data sets. While quick and efficient, this fast approach also has risks such as data loss, system downtime, and other issues.

Phased (Incremental) Migration

A phased migration strategy is an approach of migrating data in stages, usually prioritizing critical information first. This minimizes the risk of data loss or system downtime while allowing testing and validation between stages. Unfortunately, managing all these different steps can take a considerable amount of time and effort.

Parallel Migration

Parallel migration refers to running both the old and new systems simultaneously, allowing for a gradual transition. This approach works best for large organizations that need to minimize downtime or data loss while maintaining consistency between both systems. However, maintaining consistency may prove challenging in such cases.

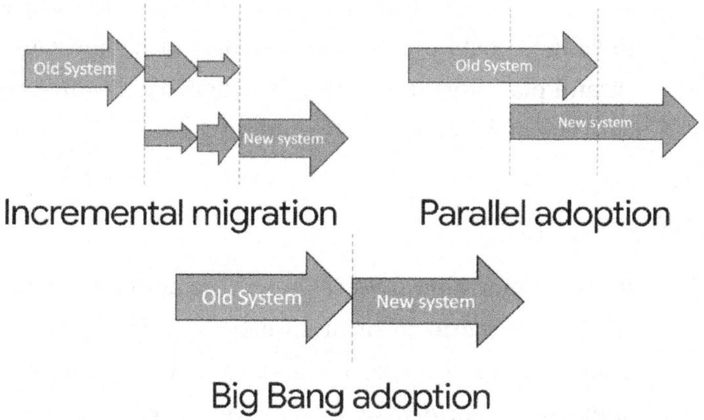

Hybrid Migration

Hybrid migration refers to a combination of multiple migration strategies, typically phased and parallel. This technique works best for organizations with complex data structures and multiple data sources; however, it may take more time and money than other migration techniques.

Re-Platform Migration

Re-platform migration refers to the process of migrating data onto a new platform or architecture. This approach can be advantageous for organizations looking to modernize their systems or transition to cloud-based solutions; however, maintaining compatibility with legacy systems may prove challenging and necessitate extensive planning and testing in order to be successful.

Re-Host Migration

Re-host migration is the process of moving applications and data to a new hosting environment without altering the application code. This method is ideal for organizations looking to reduce costs or enhance performance; however, it requires careful planning and testing in order to guarantee compatibility with the new hosting environment.

Re-Write Migration

Rewrite migration involves rewriting application code to run on a different platform or architecture. This approach is ideal for organizations wanting to modernize their systems or enhance application performance; however, it can be time-consuming and expensive due to extensive resources required and testing.

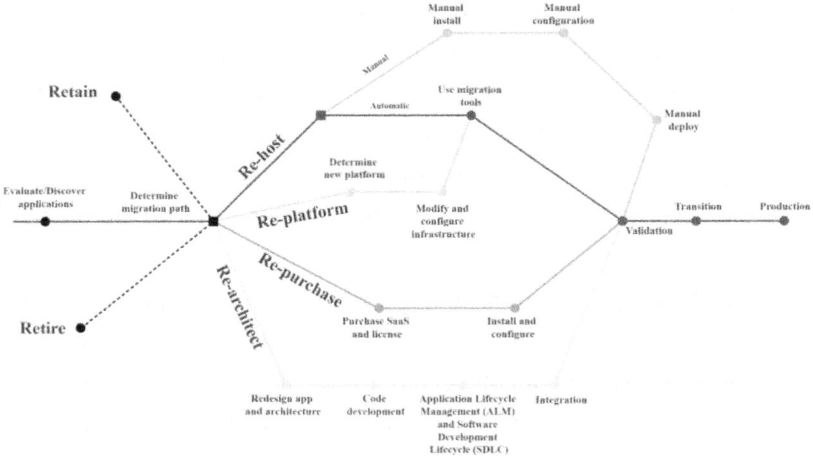

5.5 SELECTING THE RIGHT MIGRATION STRATEGY

As you begin planning for a data migration project, one of the most crucial decisions you must make is selecting an appropriate migration strategy. This decision has significant ramifications on all aspects of the project - cost, timeline and potential risks included. With so many choices available, it may seem overwhelming to select which one best fits your organization. In this section we'll address key elements to consider when making this choice and offer guidance on how to choose it correctly.

1. **Data Volume and Complexity**: When selecting a migration strategy, one of the most critical elements to take into account is the volume and complexity of your data being migrated. If there is an abundance of unstructured information that must be transferred, then considering a phased migration approach might be beneficial. A phased strategy breaks up the migration into smaller steps so it's more manageable; these smaller phases can then be completed incrementally over time.

2. **Downtime Tolerance**: Another important factor when selecting a migration strategy is how much downtime your organization can tolerate during the transition. If no downtime is acceptable during this period, parallel migration may be your best bet; this involves running both old and new systems simultaneously until all migration tasks have been completed.

3. **Data Consistency**: Maintaining data consistency throughout the migration process is paramount. If your data is highly interdependent, a cutover migration strategy may be necessary; this involves moving all the data from one system to another in one go, eliminating any chance for inconsistencies.

4. **Cost**: Cost is always a factor when selecting a migration strategy. While some options may be more costly than others, it's essential to factor in potential long-term cost savings from more complex plans. For instance, while phased migration may take longer to complete, it will ultimately result in less downtime and reduced risk.

5. **Risks and Mitigation Strategies**: Before selecting a migration strategy, it's essential to identify potential risks associated with each option and devise mitigation tactics to minimize those threats. For instance, while phased migration might be less risky than cutover migration, it requires additional resources and more time for completion.

6. **Technical Requirements**: Your organization's technical requirements will play a role in selecting an effective migration strategy. Consider aspects like hardware and software compatibility, system architecture, as well as data encryption requirements when making your selection.

Selecting the appropriate migration strategy is essential for a successful data migration project. Factors like data volume and complexity, downtime tolerance, data consistency, cost-benefit analysis, risks mitigation strategies and technical requirements should all be taken into account when making this decision. Furthermore, including stakeholders from across all areas of your organization in the decision-making process helps guarantee a smooth and successful migration.

5.6 MIGRATION PROCESS REQUIREMENTS

Migration process requirements refer to the elements that need to be taken into account when planning and executing a data migration project. These may include technical, functional, as well as non-functional aspects. Some key requirements associated with migration processes include:

Data Profiling:

Data profiling is the initial step in any migration project. This involves analyzing and understanding the data that must be moved, in order to detect any potential issues such as data quality issues, redundancy issues or inconsistencies. Once these are identified and addressed beforehand, the migration can begin.

Data Mapping:

Data mapping is another essential requirement of the migration process. Data mapping involves creating a mapping between source data and target data, helping ensure accurate migration and transformation into fit the new system. Furthermore, it helps ensure correct mapping to appropriate fields within that new system - for instance, healthcare providers may need to map patient records from an outdated electronic health record system to one using different fields and terminology.

Data Quality:

Data quality is another essential requirement of the migration process. Ensuring data quality involves identifying any potential quality issues, such as incomplete or missing data, incorrect data formats, and duplicate records. These must be addressed before beginning migration to ensure that all information is accurate, complete, and consistent. For instance, financial institutions may need to assess customer data's accuracy before migrating it onto a new system to guarantee there are no mistakes or duplicates present.

Data Validation:

Data validation is another essential step of the migration process. It involves verifying that all migrated data is accurate and complete, which can be done through various techniques like sampling or comparing it to the source data. Validating migrated information ensures it meets business requirements; for instance, retail companies need to validate all product information has been accurately transferred into an inventory management system before going live.

Testing:

Testing is an integral part of the migration process. This involves testing each step to confirm it works as intended and that data has been correctly migrated. Testing can be conducted through various techniques such as unit testing, integration testing, and user acceptance testing. Testing helps guarantee a successful migration with accurate and complete migrated information.

Documentation:

Finally, documentation is an integral component of the migration process. This involves recording all steps involved in the transfer, such as data profiling, mapping, quality assurance checks, validation procedures and testing activities. Documenting ensures that this stage can be repeated in future and any issues detected and promptly addressed.

Data Security:

Protecting sensitive and confidential data during migration is of utmost importance. For instance, government agencies must guarantee that classified information is securely transferred to a new system that adheres to security protocols.

Data Governance:

Establishing policies and procedures to guarantee data management and utilization throughout the migration process. For instance, a marketing firm might need to set up data governance procedures so customer information is used compliantly with data privacy regulations.

Meeting these requirements is critical to ensure a successful data migration process that minimizes downtime, maximizes data quality, and ensures data security and compliance.

5.7 IDENTIFYING MIGRATION RISKS & MITIGATION STRATEGIES

Before identifying migration risks, it is essential to comprehend what these threats are. Migration risks refer to any events or issues which could impede data migration processes and can originate from various sources such as:

- **Technical Issues**: Technical risks can include hardware or software failure, data corruption, or data loss during the migration process.
- **Organizational Risks**: Organizational risks arise due to poor communication, ineffective team coordination, inadequate training and resources.
- **Security Risks**: Security risks can include unauthorized access, data breaches and cyber-attacks.
- **Regulatory Concerns**: Noncompliance with legal or industry regulations such as GDPR or HIPAA can pose regulatory concerns.

Identifying Migration Risks:

Recognizing migration risks is a necessary step in developing effective mitigation strategies. Here are the steps involved in identifying migration risks:

1. **Conduct a Risk Assessment**: A risk assessment involves recognizing potential hazards and assessing their effects on the migration process. This should be conducted by an expert team knowledgeable about both the migration process itself, as well as any potential hazards.
2. **Analyze Data**: Analyzing data can help detect potential risks that may not be obvious at first glance. This analysis should take into account data quality, complexity, and dependencies.

3. **Recognize Dependencies**: Recognizing dependencies is essential for understanding how changes to one system or process may influence other related ones.
4. **Examine Past Migration Projects**: Examining past migration projects can help identify any potential risks that were encountered during previous migrations and how they were mitigated.

Mitigating Mitigation Strategies:

Once migration risks have been identified, the next step is to develop mitigation strategies. Mitigation tactics refer to actions taken to lessen potential effects of potential risks during migration. Here is a breakdown of these steps in creating mitigation plans:

1. **Prioritize Risks**: Prioritizing risks involves identifying those which have the greatest potential impact on the migration process and ranking them according to severity.
2. **Create Mitigation Plans**: Creating mitigation plans involves determining the actions that need to be taken in order to minimize potential risks. These should be specific, measurable, and achievable goals.
3. **Delegate Responsibilities**: Assigning accountability involves identifying the individuals or teams responsible for carrying out your mitigation plans.
4. **Test Mitigation Plans**: Testing mitigation plans involves verifying their effectiveness in mitigating potential risks.

Identifying migration risks and creating mitigation strategies are essential steps in ensuring a successful data migration. By anticipating potential issues and devising measures to combat them, organizations can minimize their effect on the migration process, potentially avoiding delays, financial losses, or data loss while providing for an effortless transition into the new system.

5.8 ESTABLISHING MIGRATION TIMELINES

Data migration experts know the importance of setting realistic and achievable timelines for their projects. Data migration timelines provide guidance throughout the transition, setting a duration for the move and providing an outline for everyone involved in the project.

Establishing accurate and achievable migration timelines necessitates careful planning and consideration of several critical elements, such as the complexity of the data to be migrated, its volume, the strategy chosen for migration, available resources and overall project timeline. Neglecting any one of these aspects may cause major delays and cost overruns.

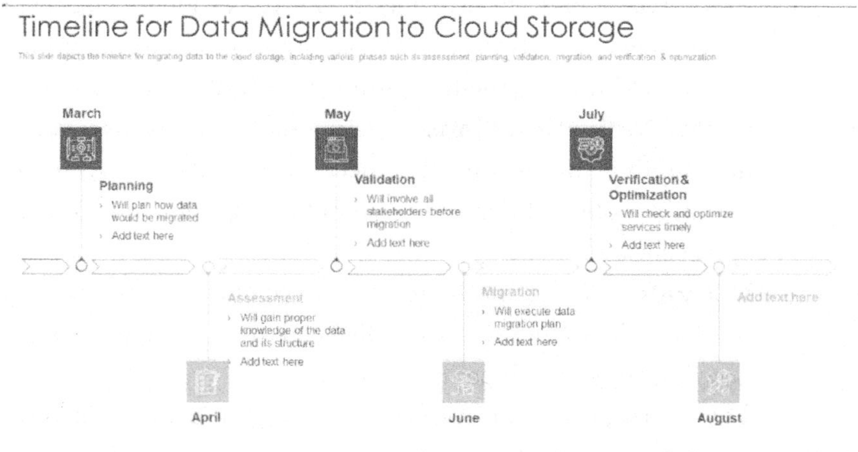

Source: https://www.slidegeeks.com/business/product/timeline-for-data-migration-to-cloud-storage-ppt-pictures-portfolio-pdf

One of the initial steps in creating migration timelines is to assess the complexity of the data being moved. This helps identify potential roadblocks during the transfer and allows the project team to formulate appropriate mitigation strategies. Data complexity is determined by factors such as type, size, quality, structure and number of sources;

more complex data may require extra time for mapping, transformation and validation which could significantly extend its timeline.

When creating migration timelines, the **volume of data** to be transferred must also be taken into account. The larger the volume, the longer it will take to transfer, map, transform and validate. Therefore, setting an achievable realistic timeframe for data transfer that takes into account available network bandwidth and speed is essential; doing so helps project teams avoid delays and optimize the migration process.

When planning a migration timeline, the **strategy** used must be taken into account. Different migration approaches exist such as the Big Bang approach, phased migration and parallel migration; each with their own advantages and drawbacks that must be carefully assessed by the project team before being adopted for their organization. For instance, migrating all data at once via the Big Bang approach versus migrating smaller chunks over time via phased approach will significantly impact timeline; hence why each option must be carefully evaluated by each team member.

Resources available are key when setting migration timelines. The project team must have enough resources to complete the migration process efficiently, such as people, infrastructure, tools and budget. They should have a comprehensive understanding of resource requirements and guarantee they are allocated appropriately in order to prevent delays or cost overruns.

Finally, the overall **project schedule** is an essential consideration when creating migration timelines. The migration plan must be integrated with the overall schedule to guarantee that it doesn't disrupt other project activities. The project team must create a detailed timeline that takes into account project milestones, deadlines and deliverables.

Establishing realistic and achievable migration timelines is essential for a successful data migration project. The project team must carefully consider several elements, such as data complexity, volume, migration strategy, resources available and project schedule. By taking an extensive approach to timeline planning, they can minimize delays and cost overruns while guaranteeing a successful migration experience.

5.9 DATA MAPPING TECHNIQUES

Data mapping is the process of recognizing the structure and format of data in one system and mapping it to its counterpart in a target system. Put simply, data mapping involves translating information from one format into another.

Data mapping is an essential step in data migration because it guarantees that the transferred information from the legacy system is accurate and complete. Without accurate mapping, transferred information may be incomplete or corrupted, leading to errors and lost data.

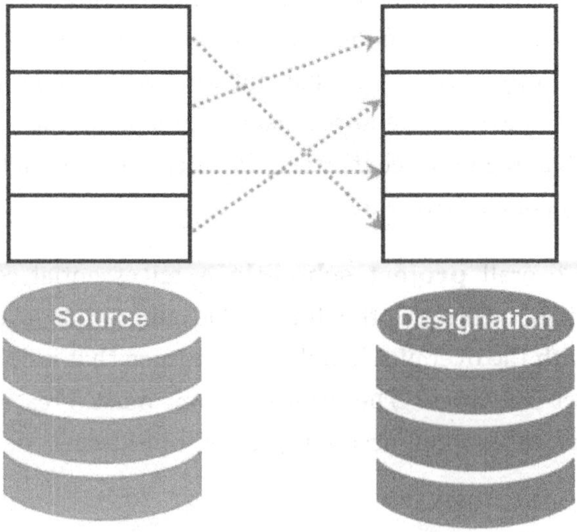

Here we present some commonly utilized data mapping techniques:

Manual Data Mapping

Manual data mapping is the simplest technique for data mapping. This involves manually recognizing the structure and format of data in one system and mapping it to its counterpart on the target one. Manual data mapping can be laborious and error prone, but may be suitable for small datasets.

Automated Data Mapping

Automated data mapping is a method that utilizes software to recognize the structure and format of data in one system and map it to its counterpart's structure and format. Automated data mapping is faster and more precise than manual mapping, making it ideal for large datasets.

Automated data mapping tools use algorithms to recognize the structure and format of data in one system and map it to its equivalent in another. They also detect and resolve conflicts within data mapping data structures, guaranteeing accuracy and completeness when transferred over.

One of the advantages of automated data mapping is its reduced risk of human error. Manual data mapping requires an analyst to manually identify and match fields, which can be time-consuming and prone to mistakes. Automated software eliminates these issues by accurately recognizing and mapping data fields in a fraction of the time it would take a human analyst.

Take the example of a large retail company migrating its customer data to a new CRM system. The legacy system stores customer information in various formats, such as first name, last name and email address; on the other hand, the new CRM requires specific formats like name, email and phone number for access.

The company utilizes automated data mapping software to recognize and match the data fields between two systems. It automatically determines that first name and last name fields from the legacy system correspond to those in the new one; additionally, it maps email from the legacy system to its equivalent field in the new one. Moreover, since phone number information is missing from the legacy system, it creates a new field for capturing it.

Automated data mapping not only guarantees accuracy of data, but it also saves time and resources during the migration process. By eliminating manual data mapping - which can be a tedious and error-prone task - automated data mapping eliminates this need for manual intervention.

Semi-Automated (Hybrid) Data Mapping

Semi-automated data mapping is a combination of manual and automated techniques for data mapping. This technique involves using automated tools to recognize the structure and format of data in one system, then manually mapping it onto its target counterpart's structure and format.

Semi-automated data mapping is useful for complex data sets where automated data mapping tools may not be able to identify the structure and format of data accurately. This technique can also help identify data mapping conflicts that may not be identified by automated data mapping tools.

5.10 BEST PRACTICES FOR DATA MAPPING

Here are some top practices for data mapping that can help make the transition successful:

Identify and Analyze Source Data

The initial step in data mapping is to identify and analyze the structure and format of data in the source system. This should include recognizing data types, relationships, and dependencies. Understanding this original information can help resolve potential data mapping conflicts and guarantee that transferred information is accurate and complete.

Define Target Data Structure

The next step is to define the structure and format of data in the target system. Doing this helps guarantee that transferred information is accurate and complete. The target data structure should be designed according to business needs, including data types, relationships between data elements, and dependencies between them.

Map Source Data to Target Data Structure

The next step is to map the structure and format of data in one system to that in the target system. This mapping should be done according to business requirements so that transferred information is accurate and complete.

Validate Data Mapping

The last step in data mapping is validation. This involves testing the mapping to guarantee it's accurate and complete before being transferred into a new system. Validation should include looking for conflicts, dependencies, and data relationships in the map itself.

5.11 DATA TRANSFORMATION TECHNIQUES

Data transformation techniques are an integral component of data migration projects, helping organizations efficiently manage the transition of data from one system to the other. Data transformation refers to any process or procedure that alters existing data in order to meet specific requirements in a target system. It involves several techniques which guarantee data quality, accuracy and consistency during migration. In this section we will explore in-depth the various data transformation techniques commonly utilized during data migration projects.

Data Transformation Rules:

Data transformation rules define the logic used to transfer data from one system to another. They guarantee accuracy and consistency in data transformation, regardless of its complexities or simplicity. Data transformation rules may include calculations, conversions, validations, and formatting as necessary - all with one goal in mind: consistent transformation according to requirements within a target system.

Data Cleansing:

Data cleansing is the process of detecting and correcting errors, inconsistencies, and inaccurate data. It's vital for ensuring data quality during migration. Data cleansing typically entails several techniques such as data profiling, standardization, parsing, and enrichment to detect missing values, incorrect values, and inconsistent data. Ultimately, the goal of data cleansing is to guarantee accurateness, consistency, and high-quality outcomes from your analytics.

Data Validation:

Data validation is the process of verifying data accuracy, completeness and consistency. This includes techniques like data profiling, standardization and cleansing to guarantee it meets business rules and requirements in a target system. Ultimately, data validation aims to guarantee high-quality information which can be relied upon for decision making.

Data Enrichment:

Data enrichment is the process of improving data by adding additional information. This involves techniques like data profiling, parsing and validation. Data enrichment helps ensure accuracy and completeness in the collected information by filling in missing details, correcting inaccurate info or adding context. Ultimately, the goal is to enhance quality and make data more useful for decision-making.

Data transformation techniques are an essential aspect of data migration projects that ensure data quality, accuracy, and consistency. These techniques involve several processes such as data mapping, data transformation rules, data cleansing, data validation, and data enrichment. Implementing these techniques helps organizations to efficiently manage the process of transforming data from the source system to the target system. It is important to select the right data transformation techniques based on the specific requirements of the target system and the nature of the data being migrated.

5.12 SUMMARY

- An understanding of Data Migration Strategy is paramount for any successful data migration project.

- It's necessary to consider all phases of the migration process - assessment, planning, execution, and validation - in order to guarantee a seamless transfer.

- Migration strategies come in various forms, such as big bang, phased, and parallel; each with its own advantages and drawbacks.

- Selecting the optimal migration strategy requires considerations such as data complexity, available resources and desired timeline for the project.

- Success of any migration project requires specifying requirements, such as data mapping and transformation techniques.

- Identification and mitigation of migration risks and development of mitigation strategies can help minimize disruptions while protecting data security and integrity.

- Setting a timeline for the migration project can help keep it on track and guarantee its completion within the specified timeline.

- Effective data mapping techniques guarantee accurate data transfer from source system to target system.

- Data transformation techniques can help guarantee data is formatted correctly and compatible with the target system, reducing the likelihood of errors or data loss.

5.13 TEST YOUR KNOWLEDGE

I. **What is the primary purpose of developing a migration strategy?**

a) To identify data quality issues
b) To define data transformation techniques
c) To ensure data readiness for migration
d) To assess migration risks

II. What are the three main phases of data migration?

a) Pre-migration, post-migration, and migration review
b) Planning, execution, and testing
c) Discovery, analysis, and design
d) Data assessment, data mapping, and data transformation

III. What is the purpose of selecting the right migration strategy?

a) To minimize the time required for data migration
b) To reduce migration risks and minimize the impact on business operations
c) To ensure that all data is migrated accurately and completely
d) To select the most cost-effective migration strategy

IV. What are the three types of migration strategies?

a) Big Bang, phased, and parallel
b) Full, partial, and incremental
c) Automated, semi-automated, and manual
d) Cloud-based, on-premise, and hybrid

V. What is the purpose of identifying migration risks?

a) To ensure that all data is migrated accurately and completely
b) To reduce migration costs
c) To identify potential issues that may impact the migration process
d) To define migration timelines

VI. What is the purpose of establishing migration timelines?

a) To ensure that all data is migrated accurately and completely

b) To define data transformation techniques
c) To minimize migration risks
d) To ensure that the migration process is completed within the required timeframe

VII. What are the two main data mapping techniques?

a) Manual mapping and automated mapping
b) Logical mapping and physical mapping
c) Source-to-target mapping and entity-level mapping
d) Flat-file mapping and hierarchical mapping

VIII. What is the primary purpose of data transformation techniques?

a) To identify data quality issues
b) To define migration risks
c) To ensure data readiness for migration
d) To ensure that data is transformed correctly during migration

IX. Which of the following is a best practice for developing a migration strategy?

a) Starting the migration process before identifying migration risks
b) Using a manual migration strategy to ensure accuracy
c) Defining clear migration timelines and milestones
d) Selecting a migration strategy based solely on cost

X. What is the purpose of migration process requirements?

a) To ensure that data is transformed correctly during migration
b) To define clear migration timelines and milestones
c) To identify potential issues that may impact the migration process
d) To minimize migration risks

5.14 ANSWERS

I. Answer: c) To ensure data readiness for migration
II. Answer: b) Planning, execution, and testing
III. Answer: b) To reduce migration risks and minimize the impact on business operations
IV. Answer: a) Big Bang, phased, and parallel
V. Answer: c) To identify potential issues that may impact the migration process
VI. Answer: d) To ensure that the migration process is completed within the required timeframe
VII. Answer: c) Source-to-target mapping and entity-level mapping
VIII. Answer: d) To ensure that data is transformed correctly during migration
IX. Answer: c) Defining clear migration timelines and milestones
X. Answer: d) To minimize migration risks

06

6 CREATING A PROJECT PLAN

Establishing a Project Plan is essential for any data migration project. It involves crafting an itinerary to guarantee its successful completion. A project plan is a comprehensive document that outlines all activities, milestones, timelines and resources necessary for its successful completion. It serves as a roadmap that directs the project team in achieving their data migration objectives. In this chapter, we'll cover the essential elements of creating a project plan, such as its importance, its various phases, and best practices for data migration projects. By the end of this chapter, you will have a comprehensive understanding of project planning tools and techniques, such as Gantt charts, work breakdown structures, and risk management strategies. With your newfound insight into creating an effective data migration project plan, you'll have all the pieces necessary for success.

6.1 IMPORTANCE OF CREATING A PROJECT PLAN

Experts in data migration must appreciate the significance of creating a project plan. This essential document lays out all of the steps and tasks necessary for successful data migration, acting as a guide for both project managers and stakeholders involved. It serves as an essential blueprint throughout every stage of the migration process.

Producing a project plan may seem like an overwhelming and time-consuming task, but it is an essential element for guaranteeing the success of any data migration project. It provides structure to the project and helps identify potential obstacles that may occur along the way. In this section, we will explore why creating a plan is so important and outline key elements that should be included within it.

Provides Clarity for the Project

A project plan lays out the objectives, timelines and goals of a data migration project. It serves as an outline for all parties involved and serves as an efficient roadmap. Without one, project teams may struggle to understand what's expected of them and how best to reach their desired outcomes. A project plan provides clarity throughout all stages of execution.

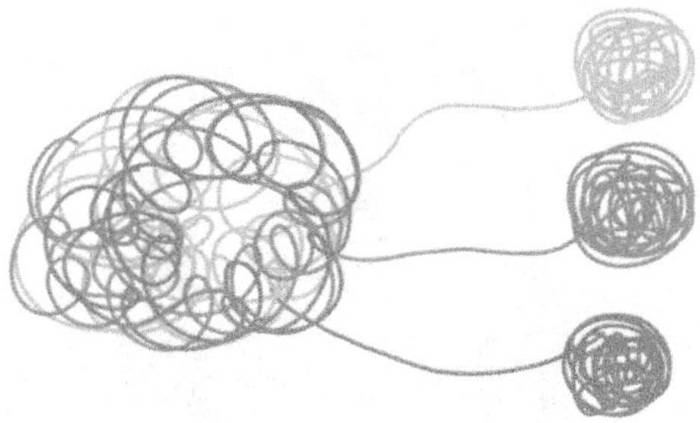

A Project Plan Fosters Project Efficiency

A project plan helps guarantee that projects are finished within their allocated timeline and budget. It outlines resources needed, timelines for each task, and a critical path for the undertaking. This helps identify any potential delays that could impact progress as well as providing a framework to manage it efficiently.

Enables Effective Communication

A project plan allows project managers and stakeholders to have effective communication. This enables everyone involved in the process to understand its progress as well as any potential issues that might arise. Furthermore, having a common document ensures all parties are working towards an agreed upon objective.

Identifies Potential Risks

One of the key advantages of creating a project plan is its ability to detect potential hazards that could negatively affect the project. By anticipating these issues ahead of time, project managers can implement mitigation strategies and minimize their effects on the undertaking - thus guaranteeing completion on schedule and within budget. This guarantees the project stays on track for success!

Establishes a Foundation for Future Projects

A well-designed project plan can serve as the blueprint for future data migration initiatives. By documenting lessons learned and best practices, project managers can use it as a template, making sure their organization is better equipped to manage such future tasks efficiently.

6.2 STEPS IN CREATING A PROJECT PLAN

Crafting a project plan is an integral step in the data migration process. This roadmap directs the team towards successful completion of the project, outlining goals, objectives, timelines, tasks and resources necessary for completion. A well-written plan can keep everyone on track while guaranteeing project success.

In this section, we'll outline the steps involved in crafting a project plan for data migration. We'll address essential components of such an undertaking and offer tips on creating one that succeeds.

Step 1: Define the Project Scope

Creating a project plan begins with delineating its scope. This involves identifying specific objectives and goals for the project. Understanding this information before starting planning ensures that everyone involved stays focused on what needs to be achieved during execution of their tasks. This also keeps everyone on track during meetings or other crucial moments during the undertaking.

Step 2: Define Project Stakeholders

Determining project stakeholders is an integral step in creating a project plan. This involves recognizing individuals, groups, or organizations with an interest or potential impact on the success of the project. Stakeholders could include project sponsors, end-users, project managers, data migration specialists - everyone who must be involved! It's crucial to include stakeholders throughout this planning phase so their needs and expectations are taken into consideration.

Step 3: Construct a Project Schedule

Creating an efficient project schedule is paramount for success. This involves creating an outline for the job and outlining specific milestones and deadlines. The plan should be realistic, giving teams enough time to finish tasks on schedule. Furthermore, flexibility should be built-in so unforeseen circumstances can be accommodated during execution of the plan.

Step 4: Define Project Tasks

Defining project tasks requires identifying the specific actions needed to meet goals and objectives. This entails breaking down the project into smaller, manageable pieces which can be assigned to team members. Establishing project tasks helps guarantee that everything is finished in an organized and structured way.

Step 5: Allocate Resources

Allocating resources is an integral step in creating a project plan. This involves identifying the necessary resources for each task, such as personnel, equipment and software. Ensure your team has access to all these elements so they can successfully finish their job.

Step 6: Define Project Deliverables

Determining project deliverables is the act of identifying the specific outputs that will be produced as a result of the effort. This could include reports, data models and system documentation. Defining these deliverables ensures everyone on the team understands their responsibilities and provides an objective measure for project success.

Step 7: Acknowledge Project Risks

Identifying project risks involves recognizing potential issues that could impede its progress, such as technical difficulties, data loss concerns and project delays. By anticipating potential issues ahead of time, teams are better able to prepare themselves to tackle them head-on and avoid delays or setbacks during the transition.

Step 8: Craft a Communication Plan

Crafting an effective communication plan is critical in creating any project plan. This involves formulating strategies for communicating with project sponsors, end-users, and team members alike. With such an arrangement in place, everyone involved in the project can stay abreast of progress on the project while offering feedback when needed.

6.3 DEFINING THE SCOPE OF THE PROJECT

Before beginning a data migration project, one of the most critical tasks is defining its scope. This step can make or break success of the entire project. When crafting scope definition, consider objectives, goals and deliverables; resources needed; constraints or assumptions that will affect progress; all these should all be taken into consideration when starting out on this journey.

The scope of a data migration project depends on the size and complexity of your organization as well as the data involved. However, there are some standard steps that can be used to define this scope.

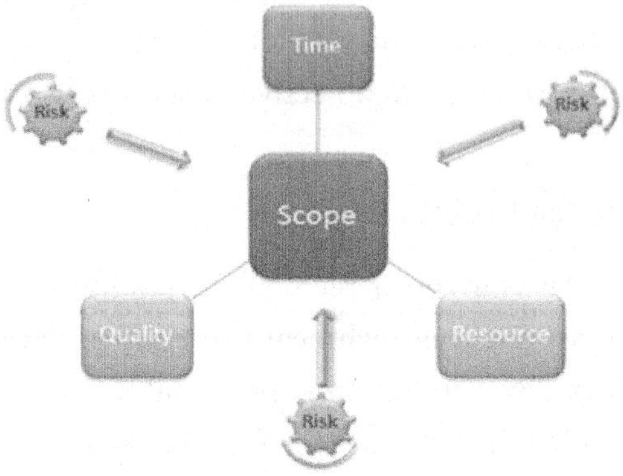

Step 1: Define Project Objectives

The initial step in defining a project scope is to clearly define its objectives. This involves answering "What do we hope to accomplish with this project?" These should be specific, measurable, achievable, relevant and time-bound (SMART). Furthermore, they should align with overall organizational goals.

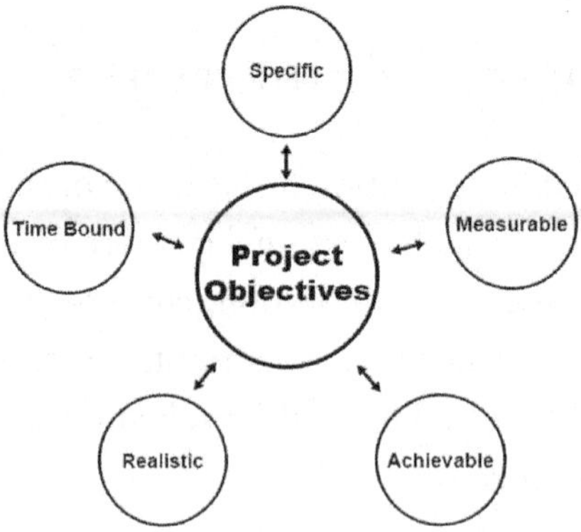

Step 2: Define the Project Deliverables

Once project objectives have been defined, the next step is to identify its deliverables. Deliverables refer to tangible outcomes produced by a project; in the case of data migration projects this might include creating a new data model, updating data maps, and successfully migrating data sets. It's essential that these deliverables be identified early in the process so they can be measured for success later on.

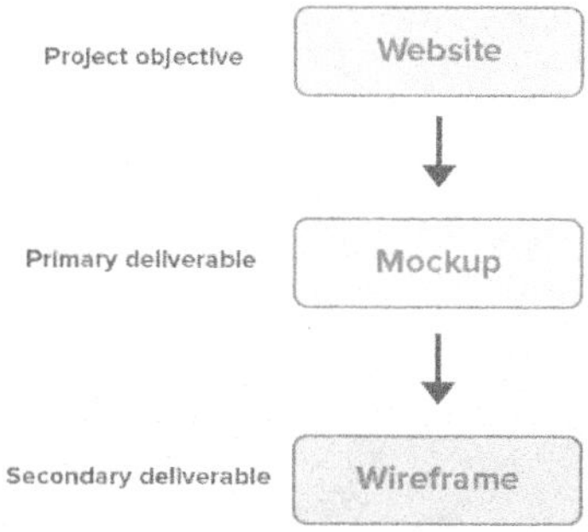

Step 3: Define Project Constraints

Project constraints are those factors which restrict a team's capacity to finish within specified scope, timeframes and budget. These could include limited resources, tight timelines or technical obstacles. It is critical to identify these obstacles early in the process so they can be addressed and mitigated as necessary.

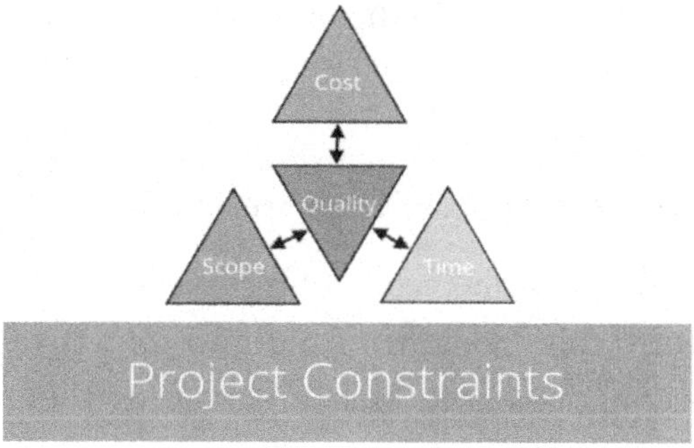

Step 4: Recognize Project Assumptions

Project assumptions are those factors which must be true in order for a project to succeed. These could include things like resource availability or software system compatibility. It's essential to identify these assumptions early on so they can be validated or disproven.

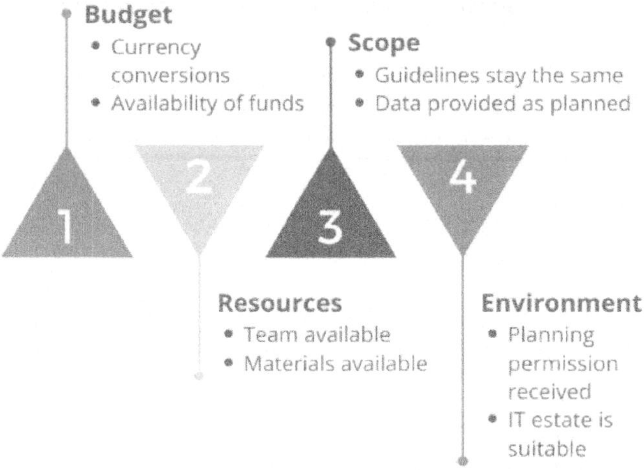

Source: https://rebelsguidetopm.com/project-assessments/

Step 5: Assess Project Risks

Every project, even data migration projects, involves risks. These could include data loss, system outage or security breaches. It's essential to recognize these hazards early in the process so they can be assessed and managed appropriately.

Step 6: Construct the Project Scope Statement

The last step in defining the scope of a project is creating the project scope statement. This document outlines objectives, deliverables, constraints, assumptions and risks related to the project; it also delineates boundaries for success and sets expectations. Serving as guidance throughout the duration of the project, this scope statement helps keep everyone on track and within budget.

Here is an example project statement:

Product Scope Description

Model T is a low-cost electric vehicle with four doors, relatively high mileage, great durability, and a low-priced interior that can be upgraded. Gas-based cars are increasing pollution, and fuel is getting expensive. Low-cost electric cars will solve these problems and help protect the environment.

Acceptance Criteria

1. The Car should run to 300km on a full charge battery
2. Low service costs
3. Low financing options from all major institutions
4. Upgrade options
5. High durability
6. Two-year warranty for external and internal parts
7. The car should achieve 110 km/hours within 10 seconds

Deliverables

1. One electric vehicle
2. One full car battery warranty
3. One spare wheel
4. One front airbag
5. Back and front leather seat covers

Project Exclusions

1. Free spare battery
2. Free charging cables
3. Extra coolant or any spare parts
4. No insurance policies
5. No free service

Constraints

1. COVID-19 may affect demand
2. The car should be launched by fall 2021
3. Budget
4. Time/resources/personnel limitations
5. Project team members will be backfilled as necessary
6. Funding for recommended and approved changes will be provided

Assumptions

1. All raw materials will be available
2. All semiconductor chips will be available
3. Permanent staff will not be reassigned
4. Funding will be available when required

Source: https://pmstudycircle.com/project-scope-statement/

6.4 SETTING PROJECT OBJECTIVES AND GOALS

Setting project objectives and goals is of the utmost importance. Without clear direction, a project may lose focus, leading to wasted time and resources. This is especially true for data migration initiatives which typically involve large amounts of information with high degrees of complexity.

Establishing clear objectives and goals helps guarantee everyone involved in the project understands what is expected of them and what the end goal is. Doing this can prevent misunderstandings or miscommunications which could quickly stall progress before it even gets off the ground.

Setting project objectives and goals helps guarantee that the project is in line with the organization's overall strategy and goals. Doing this guarantees the project adds value to the organization while furthering its broader objectives.

Setting Project Objectives and Goals

Defining project objectives and goals requires collaboration among multiple parties, such as project managers, business owners, IT staff, and other key players. To guarantee the process is successful and results in concrete objectives with achievable deadlines:

1. **Define the Project Scope**: Before setting objectives and goals for the venture, it is essential to define its scope. This involves outlining what will and won't be included in the undertaking, as well as any limitations or constraints that could potentially hinder progress.

2. **Identify Key Stakeholders**: Determine who are the key stakeholders involved in the project and ensure they are included in setting objectives and goals.

3. **Conduct a Needs Assessment**: Perform a needs assessment to identify the key requirements and priorities of project stakeholders. Doing this helps guarantee that objectives and goals are in sync with organizational requirements.

4. **Create SMART objectives**: Specific, Measurable, Attainable, Relevant and Timebound goals are the cornerstone of any successful project. They offer a concise yet precise description of what the initiative aims to accomplish and enable progress measurement and tracking throughout its duration.

5. **Align Project Objectives with Strategic Objectives**: Make sure the project objectives and goals are in sync with the organization's overall strategic targets and objectives. Doing this can guarantee that the project adds value to the organization while aiding in the pursuit of its broader objectives.

6. **Communicate Objectives and Goals**: Once project objectives and goals have been defined, it is essential to convey them clearly to all stakeholders. Doing this will guarantee everyone understands what is expected of them and the outcomes the project seeks to reach.

6.5 IDENTIFYING & ALLOCATING PROJECT RESOURCES

Data migration projects require resource allocation to guarantee success. Resources refer to people, technology and other assets needed for execution of the plan; identifying and allocating these resources is an essential step in planning as it determines project scope, timeline and budget. This section will discuss why identifying project resources is so important, as well as different types of resources commonly employed in data migration initiatives.

Identification and allocating project resources is paramount to the success of any data migration project. Without adequate funding, a venture may face delays, increased expenses, or even complete failure. Appropriate resource allocation guarantees efficient execution while managing risks to minimize their effects.

Types of Project Resources:

Resources needed for data migration projects typically include:

1. **Personnel**: Personnel responsible for carrying out the project plan, such as project managers, data analysts and developers.

2. **Hardware**: All physical equipment necessary to finish the project, such as servers, storage devices and networking gear.

3. **Software**: Data migration tools and applications like data integration software or database management systems must also be utilized.

4. **Data**: Data to be transferred from the source system to the target system.

5. **Facilities**: Physical spaces required for project activities such as office space, meeting rooms and testing environments.

6. **Budget**: All financial resources necessary for completion of the project, such as personnel costs, hardware/software purchases and facilities use.

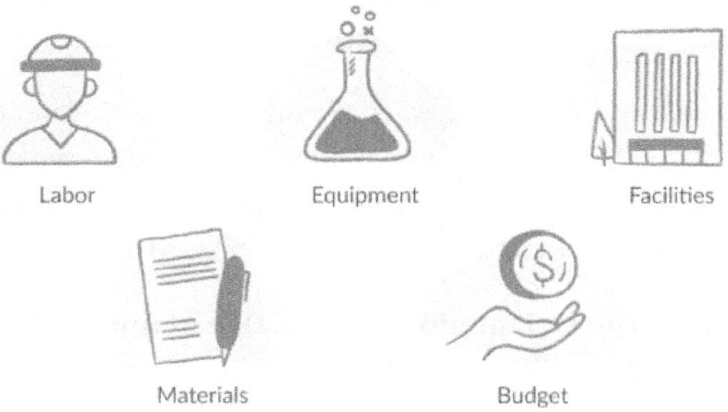

Identifying Project Resources:

To properly identify project resources, it is necessary to have an extensive understanding of the project objectives, scope and timeline. This info helps determine what types and quantities of resources will be needed for success. Typically, the project manager is accountable for identifying and allocating these funds with input from stakeholders as well as team members.

Allocating Project Resources:

Once project resources have been identified, they must be allocated efficiently in order for the project to run smoothly. Allocation involves assigning resources to specific tasks and activities while setting a timeline for completing those tasks. The allocation process should be managed carefully in order to make use of resources efficiently and ensure the project stays on track.

6.6 CREATING A PROJECT TIMELINE

A project timeline is an essential tool in project planning that acts as a roadmap to guide the team through each stage of the project. It displays key milestones, tasks and activities that must be accomplished within an established time frame to ensure success.

A timeline is typically created during the project planning phase and updated throughout its execution as new information becomes available. A well-crafted and realistic project timeline is essential to any successful project, offering several advantages such as:

1. **Project Timeline Aids in Determining Milestones and Their Timing:** A project timeline can assist in pinpointing critical milestones and their associated timelines. It gives everyone a clear idea of when each phase should be completed, which allows you to track its progress over time.

2. **Facilitates effective risk management**: A project timeline assists in recognizing potential risks that could delay or cancel out a milestone. With this data, managers can devise risk mitigation strategies and minimize their effect on the schedule.

3. **Determining Project Completion on Schedule**: A project timeline helps define realistic deadlines and guarantees the project is finished within its designated timeline. By breaking down the project into smaller tasks, it allows the team to focus on completing individual tasks within their designated deadline.

4. **Improved Communication**: The timeline serves as a visual aid that can be utilized to convey project status and progress to stakeholders. It guarantees everyone is on the same page regarding timelines and expectations, which reduces the chance for misunderstandings or delays.

5. **Facilitating Resource Management**: A well-crafted timeline allows project managers to allocate resources

according to the timeline's demands. It helps identify resource gaps and guarantees the right people are assigned to the right tasks at the correct time.

Steps in Crafting a Project Timeline

1. **Define the Project Scope**: The initial step in crafting a project timeline is to define its scope. This includes specifying objectives, deliverables and desired outcomes. Analyze Cost Effectiveness and Schedule Accuracy.

2. **Break Down the Project into Tasks**: The next step in any successful project is to break it down into smaller, achievable objectives. Doing this helps identify dependencies between tasks and create a critical path that must be followed for timely completion of all tasks.

3. **Estimate Task Durations:** Once you've divided your project into smaller tasks, the next step is to estimate their durations. This can be done based on past experience, expert judgment or other project management tools.

4. **Recognize Milestones**: Milestones are significant events or accomplishments that take place during a project's lifespan. They serve as checkpoints for the team to monitor progress towards achieving objectives. Here is an example of a milestone:

5. **Establish Dependencies**: The next step is to determine the dependencies between tasks. This helps decide which ones must be finished before others can begin, and which can be completed simultaneously.

6. **Create a Gantt Chart**: A Gantt chart is an illustrated timeline that displays tasks, milestones and dependencies on it. It helps project managers detect potential delays and adjust their plans accordingly. Here is an example of one type of Gantt Chart:

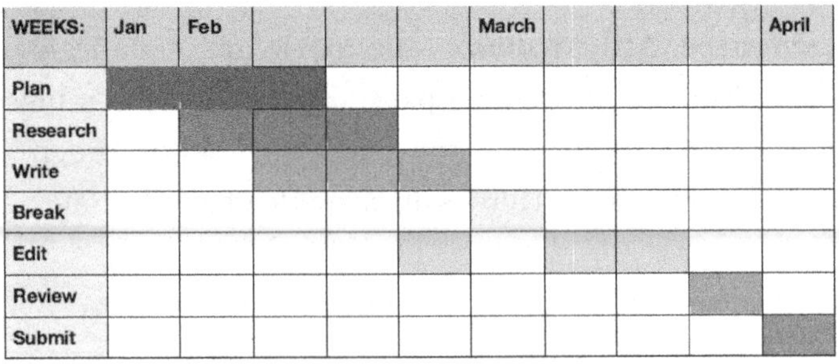

7. **Allocate Resources**: The final step in creating a project timeline is assigning resources to each task. This involves identifying the people, equipment and materials needed to finish each task within the defined timeline.

6.7 ESTABLISHING KEY PERFORMANCE INDICATORS

Data migration is a complex undertaking that necessitates meticulous planning and execution to guarantee its success. To effectively manage and monitor the progress of a data migration project, key performance indicators (KPIs) should be established - these are quantifiable metrics used to judge the success of any venture or process.

Establishing KPIs for data migration projects is essential to guarantee their success. KPIs offer a way of measuring progress and assessing whether the project is on track to meet its objectives. They also help project managers detect potential issues early on, allowing them to be addressed before becoming major issues. KPIs can be utilized to evaluate individual team members' performance as well as the project overall, helping identify areas for improvement.

Identifying and Define KPIs:

When setting KPIs for a data migration project, the initial step is to identify key areas that need measuring. This will vary depending on the specific goals and objectives of the undertaking; however, common metrics associated with data migration projects include:

A. **Data Accuracy**: This KPI measures the accuracy of data being migrated. It can be assessed by comparing the information in the source system with that in the target system.

B. **Data Completeness**: This KPI measures the completeness of data being migrated. It can be measured by comparing the number of records in the source system to those in the target system.

C. **Data Consistency**: This KPI measures the consistency of data being migrated. It can be assessed by comparing data in one system to that in its target system.

D. **Data Migration Timelines**: This KPI measures the timeline for completing a data migration project by comparing the planned timeline with actual timeline.

E. **Data Migration Costs**: This KPI measures the expense associated with data migration projects by comparing the planned budget to actual expenditure.

Once the key areas have been identified, the next step is to define specific metrics to measure them. For instance, for data accuracy KPI, one metric might be the percentage of data that has been correctly migrated. It's essential that these measurements be objective, measurable and directly tied to the key areas being assessed.

6.8 RISK MANAGEMENT PLANNING

Risk management planning is essential in data migration projects as it helps organizations identify potential risks and take proactive measures to minimize them. By recognizing risks early in the project lifecycle, organizations can reduce their impact on timelines, budgets, and overall success.

Risk management planning also assists organizations in prioritizing risks based on their potential impact and likelihood. By prioritizing threats, organizations can allocate resources more efficiently and take actions to mitigate the most critical threats first.

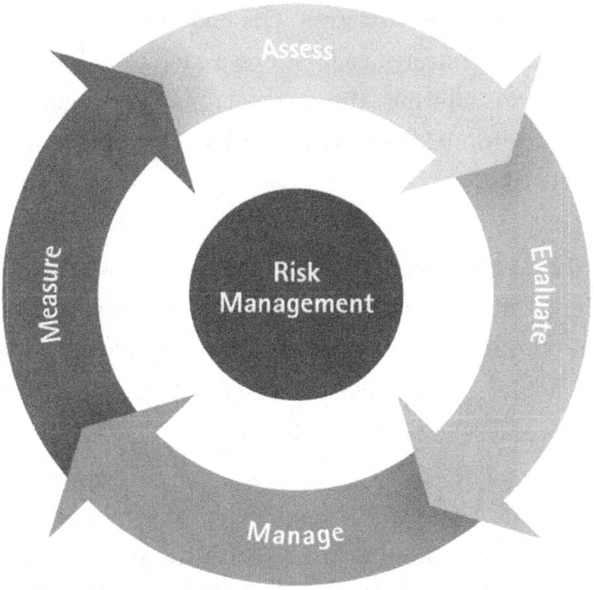

Steps in Risk Management Planning:

1. **Identify Potential Risks**: The initial step in risk management planning is to identify potential hazards. This involves reviewing the project plan and noting any areas where risks could arise. It could also involve brainstorming sessions between project team members and stakeholders to identify potential issues.

2. **Assess Risks**: Once potential risks have been identified, the next step is to assess each one's impact and likelihood. This involves considering potential outcomes of each risk and estimating its likelihood.

3. **Prioritize Risks**: After assessing risks, the next step is to prioritize them according to their impact and likelihood. Prioritizing risks involves deciding which threats are most critical for project success and require immediate attention.

4. **Craft Mitigation Strategies**: Once identified, the next step is to devise mitigation strategies to address each risk. Potential solutions could include altering the project plan or scope change, or allocating additional resources in order to minimize adverse effects from such risks.

5. **Delegate Accountability**: Once mitigation strategies have been identified, it is time to assign responsibility for their implementation. This involves designating project team members or stakeholders with specific responsibilities in order to take actions necessary to reduce each risk.

6. **Monitor and Assess**: The final step in risk management planning is to continuously monitor and assess the project's progress. This entails identifying risks, assessing their implications for the project, and making necessary adjustments to mitigation strategies as needed.

6.9 COMMUNICATION PLANNING

Effective communication is fundamental for the success of any project, and data migration is no exception. Communication planning helps guarantee all stakeholders are informed about project goals, timelines and risks so they can make informed decisions. A well-thought out communication strategy also builds trust and engagement between participants while creating an atmosphere of openness and collaboration.

Components of a Communication Plan:

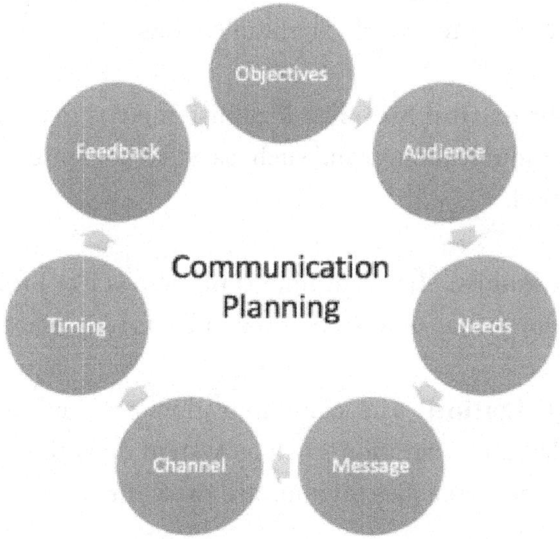

A communication plan typically comprises the following components:

1. **Stakeholder Analysis**: Determining all stakeholders involved in the project and their communication needs and preferences. This includes both internal stakeholders such as project team members or senior management, as well as external ones like customers or vendors.

2. **Communication Objectives**: Define the specific objectives of your communication plan, such as informing stakeholders of project milestones, risks, and changes.

3. **Communication Channels (Methods):** Determining the most efficient communication channels for each stakeholder group, such as email, phone or in-person meetings, depends on their objectives.

4. **Communication Schedule (Frequency):** Create a regular communication plan with stakeholders that outlines the frequency, format, and content of all communications.

5. **Communication Metrics**: Establish metrics to gauge the success of a communication plan, such as stakeholder engagement and satisfaction.

6. **Key Messages**: Highlight important details like project timelines, system downtime, or any potential impacts on patient care.

7. **Risk Mitigation**: The communication plan addresses potential risks related to migration, such as system downtime and data loss, and provides strategies to minimize these threats.

Here is an example of these components:

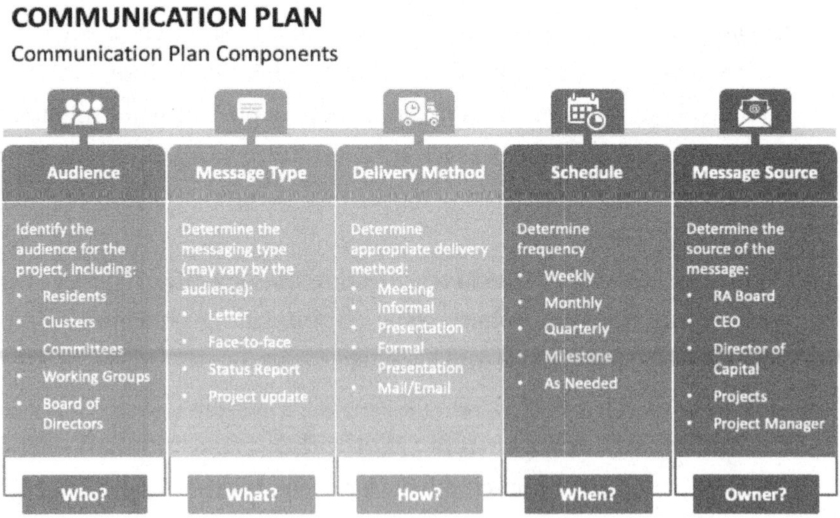

Effective Communication Strategies:

Effective communication strategies are paramount to the success of a data migration project. Some key tactics include:

1. **Customizing Communications:** Different stakeholders have differing communication needs and preferences, so it is essential to tailor messages according to their specific requirements. For instance, senior management may prefer high-level summaries and key metrics while project team members need more specific details about project tasks and timelines.

2. **Two-Way Communication**: Effective communication is a two-way affair that involves active listening and feedback. Encouraging stakeholders to provide input and ask questions can help identify and address problems before they escalate into larger problems.

3. **Consistent Messaging**: Communicating consistently is essential to reduce confusion and guarantee that all stakeholders have an accurate view of project goals and progress.

4. **Timeliness**: Timely communication is paramount in data migration, as delays or issues can have significant repercussions on project timelines and budgets.

5. **Communication Culture**: Fostering an atmosphere of effective communication within your project team can help foster trust, collaboration, and accountability.

Effective communication is a fundamental element of any successful data migration project. Communication planning plays a significant role in guaranteeing all stakeholders are informed, engaged and empowered to make informed decisions. By tailoring communications according to stakeholders' needs, employing effective communication tactics and cultivating an environment of open and transparent dialogue, data migration projects can reach their objectives on schedule and within budget.

6.10 PROJECT MONITORING AND CONTROL

Project Monitoring and Control are essential elements to the success of any data migration project. They enable the team to keep tabs on progress, adjust plans as necessary, and ensure everything runs smoothly. Here are a few reasons why project monitoring and control is so essential in data migration:

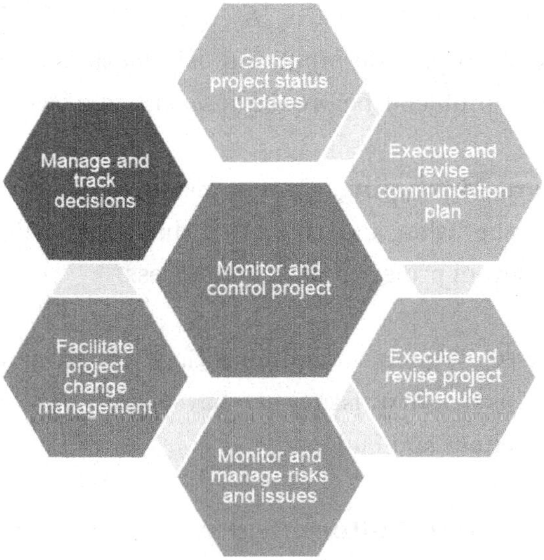

1. **Ensuring Project Success**: Effective project monitoring and control can guarantee that projects are finished on schedule, within budget, and according to quality expectations.

2. **Alerts the Project Team of Potential Issues**: By continuously monitoring the project, the project team can detect potential issues

early and take proactive measures to address them before they become bigger issues.

3. **Provides Insight for Decision Making**: Project monitoring and control provide essential data to stakeholders that can be used to make informed decisions about the project, such as when to adjust timelines or reallocate resources.

Key Components of Project Monitoring and Control:

Here are the essential elements of project monitoring and control in a data migration project:

1. **Performance Tracking**: This involves monitoring project progress against its planned path, noting any deviations from it.

2. **Risk Management**: This involves recognizing potential risks to the project and creating strategies to mitigate or manage them.

3. **Change Management**: Coordinating any modifications made to the plan or scope, along with documenting them correctly and notifying stakeholders of their effects.

4. **Communication Management**: This involves creating effective channels of communication between the project team and stakeholders, as well as guaranteeing information is shared promptly and openly.

5. **Quality Assurance**: This involves verifying that the project meets established quality standards, taking corrective actions when necessary, and effectively implementing project monitoring and control mechanisms.

Here are some tips for effectively implementing project monitoring and control in a data migration project:

1. **Define Clear Project Objectives**: Make sure all stakeholders agree upon project objectives that everyone works towards together. This ensures everyone stays focused on reaching those same results.
2. **Create a Comprehensive Project Plan**: Create an organized plan with timelines, resource allocations and key milestones.
3. **Assign Responsibilities**: Clearly define roles and responsibilities for each team member so they are understood clearly.
4. **Utilize Project Management Tools**: Utilize project management tools such as Gantt charts to monitor progress on the project and detect any deviations from plan.
5. **Host Regular Project Meetings**: Hold regular project meetings to discuss progress, identify issues, and make necessary adjustments.

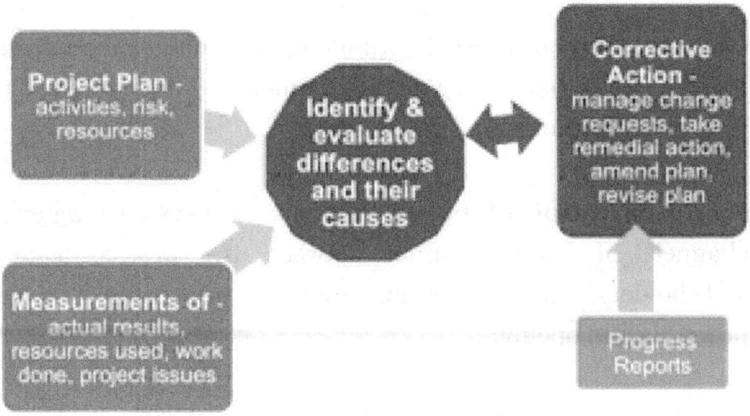

Project monitoring and control is an essential process in data migration, as it helps keep the project on track and any issues addressed promptly. Effective implementation of project monitoring and control requires clear objectives, a detailed plan, as well as effective communication between team members and stakeholders. By following these best practices, project teams can increase their chances of a successful data migration project.

6.11 SUMMARY

- A well-formulated project plan is essential for the success of any data migration effort.
- This blueprint lays out all the steps, timelines, and resources necessary for the project.
- The initial step in creating a project plan is to define its scope, including data to be migrated and objectives/goals.
- Acquiring resources is essential for making sure you have enough people, budget, and technology for the project.
- Establishing a project timeline is essential to guarantee the project is finished within its allocated timeline.
- Setting key performance indicators can help track progress and identify areas requiring corrections.
- Risk management planning is necessary to anticipate potential issues that could arise during the project and devise mitigation strategies.
- Communication planning guarantees all stakeholders are updated on project progress as well as any issues that arise.

- Project monitoring and control are essential to monitor progress against the project plan and make necessary adjustments for successful completion.

6.12 TEST YOUR KNOWLEDGE

I. What is the first step in creating a project plan?

a) Defining the scope of the project
b) Identifying project resources
c) Creating a project timeline
d) Establishing project objectives and goals

II. Why is it important to set project objectives and goals?

a) To allocate project resources effectively
b) To establish a project timeline
c) To identify project risks
d) To guide decision making and monitor progress

III. What is the purpose of identifying and allocating project resources?

a) To establish project objectives and goals
b) To define the scope of the project
c) To create a project timeline
d) To ensure that the necessary resources are available for the project

IV. What is the main purpose of creating a project timeline?

a) To define the scope of the project
b) To identify project risks
c) To allocate project resources effectively
d) To plan and manage project activities and deadlines

V. **What is a key benefit of establishing key performance indicators (KPIs)?**

a) To identify and mitigate project risks
b) To allocate project resources effectively
c) To evaluate project success and progress
d) To establish project objectives and goals

VI. **What is the main purpose of risk management planning?**

a) To allocate project resources effectively
b) To monitor and control project activities
c) To identify and mitigate project risks
d) To establish project objectives and goals

VII. **What is the purpose of communication planning?**

a) To establish project objectives and goals
b) To identify project risks
c) To allocate project resources effectively
d) To plan and manage project communications

VIII. **What is the purpose of project monitoring and control?**

a) To identify and mitigate project risks
b) To allocate project resources effectively
c) To evaluate project success and progress
d) To plan and manage project communications

IX. **What is the main purpose of a project plan?**

a) To establish project objectives and goals
b) To identify and mitigate project risks
c) To allocate project resources effectively
d) To guide project activities and monitor progress

X. What is a key benefit of project monitoring and control?

a) To establish project objectives and goals
b) To identify and mitigate project risks
c) To allocate project resources effectively
d) To evaluate project success and progress

6.13 ANSWERS

I. Answer: a) Defining the scope of the project
II. Answer: d) To guide decision making and monitor progress
III. Answer: d) To ensure that the necessary resources are available for the project
IV. Answer: d) To plan and manage project activities and deadlines
V. Answer: c) To evaluate project success and progress
VI. Answer: c) To identify and mitigate project risks
VII. Answer: d) To plan and manage project communications
VIII. Answer: c) To evaluate project success and progress
IX. Answer: d) To guide project activities and monitor progress
X. Answer: d) To evaluate project success and progress

07

7 BUDGETING AND RESOURCE ALLOCATION

Budgeting and resource allocation are two integral components of any data migration project. Planning ahead, allocating resources, and setting a budget all play an integral role in the success of the project. Effective budgeting guarantees no unexpected expenses or cost overruns, keeping the project within its allocated budget. Resource allocation ensures the right resources are available at the correct time to meet project timelines and objectives. In this chapter, we will cover the essentials of budgeting and resource allocation for data migration projects. We'll take a closer look at the steps involved in budgeting, identify the costs associated with it, and identify techniques for resource allocation. This chapter will also address the difficulties encountered when budgeting and resource allocation, as well as strategies to overcome them. By the end of it all you will have a solid grasp on how to efficiently budget and allocate resources for data migration projects.

7.1 IMPORTANCE OF BUDGETING AND RESOURCE ALLOCATION

Data migration projects can be complex and require significant resources to ensure success. Budgeting and resource allocation are essential elements of project planning that must be taken seriously. Budgeting involves estimating project costs, calculating resource requirements, and deciding how best to allocate those resources efficiently. By setting realistic budgets and allocating resources efficiently, organizations can increase their chances of completing successful data migration initiatives.

In this section, we'll examine the significance of budgeting and resource allocation in data migration projects. We'll outline key benefits associated with both, identify potential challenges organizations may experience when budgeting or allocating resources, and offer best practices for successful budgeting and resource allocation.

Key Benefits of Budgeting and Resource Allocation:

Here are some of the key benefits of budgeting and resource allocation:

Aids in Controlling Project Costs:

Effective budgeting helps control project expenses by identifying and managing costs throughout its lifespan. By having a thorough understanding of costs, organizations can avoid overruns and guarantee the project stays on track.

Ensures Resource Availability:

Effective resource allocation guarantees the right resources are available when they're needed. This includes making sure all necessary staff, equipment and materials are on hand when required for a successful project completion.

Improvements Project Planning:

Budgeting and resource allocation are essential elements of project planning. By identifying the costs and required resources, organizations can better create a timeline, set achievable objectives, and guarantee that the project stays on track.

Facilitates Decision-Making:

Effective budgeting and resource allocation provide organizations with the information necessary to make informed decisions about a project, including whether to proceed, allocating which resources for it, and managing risks effectively. This information allows organizations to make educated choices about whether to proceed with the venture.

Challenges in budgeting and allocating resources:

Budgeting and resource allocation are crucial aspects of project planning, but they can be challenging for organizations to manage efficiently. Some issues that organizations may face include:

Lack of Accurate Cost Estimates:

Estimating the costs for a data migration project can be tricky, particularly if your organization has little to no experience with similar initiatives.

Resource Constraints:

Organizations may face difficulty allocating resources for a project due to competing priorities or limited availability of resources.

Shifting Project Requirements:

Project requirements may shift, affecting costs and resource requirements as well. It becomes difficult to stay on schedule when these variables fluctuate.

Stakeholder Management:

Effective stakeholder management is essential to the success of any data migration project. But managing expectations while staying within budget and resource limits can prove challenging.

Here is an example of a project Budget template:

Project Budget Template

Summary Cost of the Project	Amount (in		Details of the Project
Total budgeted Cost during the	-		Name of the Company
Total Actual Cost during the period	-		Project Name or ID
Total Variance during the period	-		Project Lead
			Start Date

S.No.	Particulars	Material		Labor		Fixed Cost	Miscellaneous Cost	Budgeted Amount (in $)	Actual Amount (in $)
		Units	Cost per Unit	Hours	Cost per Hour				
	Task 1								
1	Subtask 1	-	-	-	-	-	-	-	-
2	Subtask 2	-	-	-	-	-	-	-	-
3	Subtask 3	-	-	-	-	-	-	-	-
4	Subtask 4	-	-	-	-	-	-	-	-
5	Subtask 5	-	-	-	-	-	-	-	-
(A)	Total Task 1					-	-	-	-

Best Practices for Effective Budgeting and Resource Allocation:

Organizations facing these obstacles must adopt best practices in budgeting and resource allocation. Some of these strategies include:

Accurately Estimate Project Costs:

In order to accurately calculate project costs, organizations should collect as much information as possible about the scope, timeline and requirements.

Create a Comprehensive Project Plan:

A comprehensive project plan helps organizations identify the resources required for success and guarantee they are available when required.

Monitor project progress:

Regularly tracking project progress allows organizations to detect potential issues early and take corrective action before they become serious issues.

Engage stakeholders:

Engaging stakeholders throughout the project helps to manage their expectations, identify issues, and guarantee that everything remains on track.

Effective budgeting and resource allocation are paramount to the success of data migration projects. By setting realistic budgets, identifying resource requirements, and allocating resources efficiently, organizations can increase their chances for successful project delivery. But in order to realize these advantages, organizations must overcome the difficulties associated with budgeting and resource management and adopt best practices for successful project management.

7.2 IDENTIFYING RESOURCES NEEDS

Data migration is a complex process that requires extensive planning, resources, and expertise for successful completion. One of the most essential elements of data migration is identifying all necessary resources needed for completion. Doing this helps guarantee that the project stays on track, is completed within budget, and meets desired outcomes. In this section we'll explain why identifying resource needs is so important and offer guidelines on how to identify them effectively.

Why Is Identifying Resource Needs Important?

Identification of resource requirements is paramount to a successful data migration project. Here are some reasons why:

1. **Accurate Budget Estimating:** Determining resource needs helps project teams accurately estimate their budget requirements. With an understanding of all necessary resources, the team can create an accurate budget that takes into account all aspects of the project.

2. **Allocate Resources Efficiently:** By identifying resource needs, the project team can allocate resources efficiently and ensure all aspects of the project are adequately covered. Doing this helps prevent over allocating resources to certain parts of the venture while neglecting other essential areas.

3. **Plan and Schedule Work Effectively**: Determining resource needs is essential for project teams when they must plan and prioritize work effectively. This becomes especially critical when working under tight deadlines; having a comprehensive understanding of required resources allows the team to plan accordingly, ensuring the project is finished within its allotted timeframe.

4. **Anticipate and Mitigate Risks**: Determining resource needs helps project teams anticipate and minimize risks related to their project. By having a clear picture of what resources are necessary, they can identify potential issues early on and take measures to address them.

5. **Enhance Quality of Work**: By accurately identifying resource needs, the project team can guarantee high-quality workmanship throughout. This is especially crucial when working on a complex data migration project where even minor mistakes have an immense impact.

Guidelines for Determining Resource Needs:

1. **Define Project Scope:** Determining a project scope is essential in accurately determining resource requirements. The team needs to have an in-depth understanding of its objectives, goals and scope in order to determine which resources are necessary to finish it successfully.

2. **Define Stakeholders**: Determining stakeholders is paramount to guarantee all aspects of a project are taken care of. Stakeholders could include project managers, IT personnel, end-users and other parties involved in the project.

3. **Define Roles and Responsibilities Accurately**: It is essential to define roles and responsibilities in order to accurately identify resource requirements. The project team needs a firm grasp on who will be accountable for what aspects of the venture in order to accurately calculate required personnel numbers.

4. **Establish Skill Sets Required**: Determining the necessary skill sets is paramount to guaranteeing that the project team has enough expertise to complete their task successfully. Therefore, they should identify these necessary abilities and guarantee they are accessible to everyone involved in the project.

5. **Determine Resource Availability**: Determining resource availability is essential to guarantee that the project team has access to all necessary resources, such as human resources, equipment, and other items needed for successful completion of the project. This includes determining whether human resources or equipment will be necessary during each stage of execution.

7.3 CREATING A RESOURCE PLAN

One of the most essential elements for a successful data migration project is resource allocation. This involves identifying necessary resources, ascertaining their availability and allocating them with maximum efficiency and effectiveness. In this section we'll look at how to create an effective resource plan as part of your overall project strategy.

Step 1: Define Project Roles and Responsibilities

The initial step in creating a resource plan is to define project roles and responsibilities. This involves identifying different tasks that need to be performed during the venture, assigning those duties to specific individuals or teams so everyone understands their role within the project and what expectations there are for them. Examples of project roles include project manager, technical lead, data analyst, data migration specialist - just to name a few!

Step 2: Define Required Resources

Once project roles have been defined, the next step is to identify the resources necessary for each function. This includes personnel, equipment, software and other resources essential to finish the project successfully. It's essential that these resources be identified early in the planning process so they are available when needed.

Step 3: Assess Resource Availability

After identifying the necessary resources, the next step is to assess their availability. This involves confirming personnel and equipment availability as well as software licenses and other resources. Conducting this assessment is essential in guaranteeing that necessary personnel and equipment are available when needed, keeping your project on schedule, and making sure all projects have enough funds when they're needed.

Step 4: Allocate Resources

Once the necessary resources have been identified and their availability assessed, the next step is to allocate them. This involves assigning specific personnel to each role, scheduling equipment usage, deciding on software licenses and other resources as necessary. Ideally, your resource allocation plan should be flexible enough to adjust according to changes in project schedule or scope.

Step 5: Monitor Resource Usage

The last step in creating a resource plan is to monitor its usage throughout the project. This involves tracking personnel, equipment, software and other resources used in order to make adjustments as necessary. Doing this helps guarantee that projects stay on schedule while using resources efficiently.

Benefits of a Resource Plan

A resource plan offers several advantages that should not be overlooked: it ensures efficiency throughout all stages of production while helping keep costs low.

Creating a resource plan is critical for any data migration project's success. Some of its advantages include:

1. Assuring that necessary resources are available when needed.
2. Minimizing resource conflicts and optimizing resource usage.
3. Minimizing the risk of project delays and overruns.
4. Improving overall project management and communication.

7.4 ESTIMATING PROJECT COSTS

Every project necessitates a significant amount of resources, and funding is one of the most essential. Therefore, accurate project cost estimation is essential to guarantee projects stay within budget and are completed successfully. Estimating project costs can be an intricate process that takes into account various elements. In this section we'll discuss why it's so important to estimate project costs accurately as well as provide various methods for doing so.

Estimating project costs is essential for several reasons, such as:

1. **Budget Planning**: Accurately estimating project costs allows for effective budget management. Accurately determining how much funding is necessary and where those funds should go helps determine how much funding should be allocated to the venture.

2. **Resource Allocation**: Estimating project costs allows for more accurate resource allocation, as it helps determine which resources are necessary and how much of each type is necessary.

3. **Risk Management**: Estimating project costs can help project managers identify potential risks associated with the venture. They're able to recognize areas where costs may run over budget and take appropriate measures to mitigate them.

4. **Decision-Making**: Estimating project costs gives project managers the information needed to make informed decisions. Knowing how much it will cost will allow them to allocate resources efficiently, prioritize tasks and manage timelines effectively.

Methods for Estimating Project Costs:

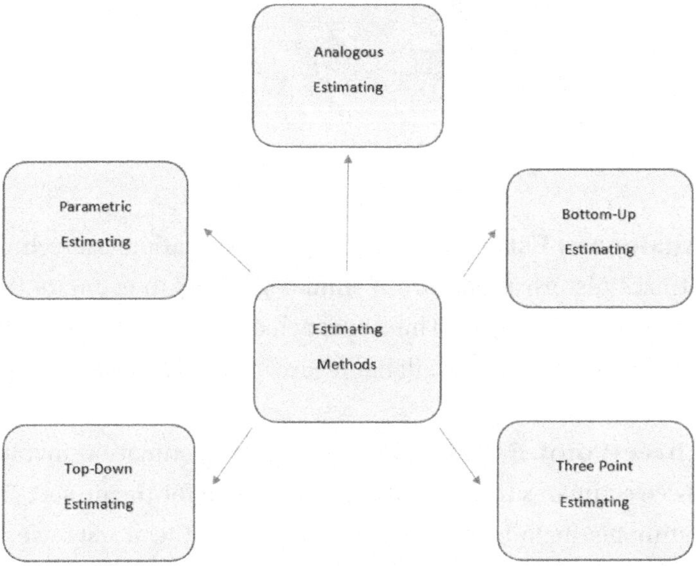

1. **Top-Down Estimation**: Top-down estimation involves estimating project costs at a high level and then breaking them down into smaller components. This method works best when there is limited information about the project and an accurate rough estimate is required.

2. **Bottom-Up Estimation**: Bottom-up estimation involves estimating project costs at a granular level and then aggregating them to arrive at an overall project expense. This method works best when detailed information about the project is available.

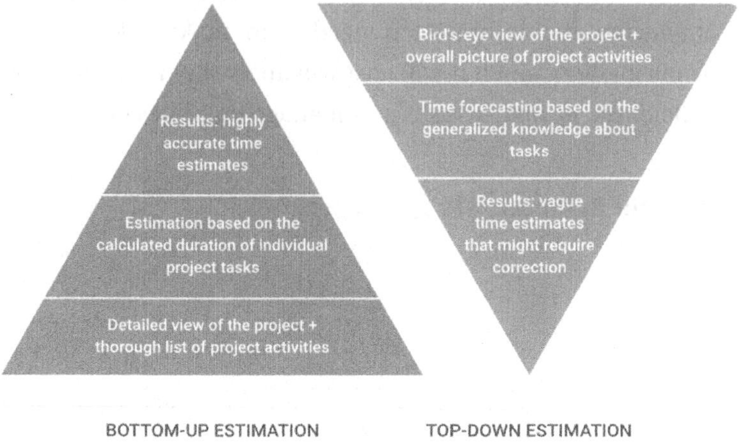

3. **Analogous Estimation**: Analogous estimation is a technique that utilizes historical data from similar projects to estimate the cost of an upcoming project. This approach can be beneficial when there is limited information available regarding a given venture.

4. **Three-Point Estimation**: Three-point estimation involves using three estimates to arrive at a cost estimate for the project. The three estimates include the best-case scenario, the worst-case scenario, and the most likely scenario. This method is useful when there is uncertainty about the project.

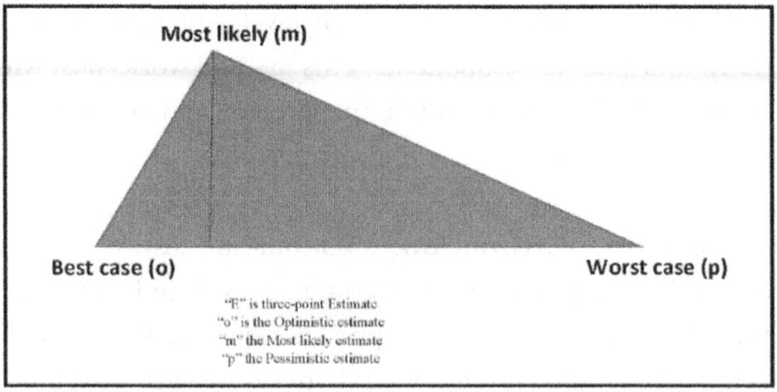

7.5 CREATING A PROJECT BUDGET

A project budget is an integral element of any successful project management plan, outlining the financial requirements and limitations. Crafting one requires identifying all resources needed for completion as well as estimating how much money is available from resources needed and where the money will go. A well-crafted budget not only keeps projects on track within available funds but also gives stakeholders a complete picture of the financial health of the project.

In this section, we will dive deep into the process of creating a project budget, outlining its essential steps and best practices.

Step 1: Define the Project Scope

Establishing a budget for any project begins with clearly delineating its scope. This includes outlining objectives, goals, deliverables and timelines. Without an accurate understanding of what needs to be achieved, estimating resources required is impossible.

Step 2: Define Resources

The next step in creating a project budget is to identify all resources necessary for its completion. This includes personnel, equipment, materials and other assets. Each resource's cost should be estimated based on its duration in the project as well as its availability and rate.

Step 3: Calculate Costs

Once resources have been identified, the next step is to estimate their costs. This includes both direct costs (e.g., salaries), indirect costs (e.g., overhead expenses, administrative fees). Break these down into more manageable components like labor hours or individual items for accurate estimates.

Step 4: Allocate Resources and Costs

Once resources and costs have been estimated, the next step is to allocate them to specific tasks or work packages. This involves estimating how much of each resource is necessary for each task as well as its associated cost. Allocating should be done in a way that meets both project objectives while making best use of available resources.

Step 5: Construct a Budget Summary

The concluding step in creating a project budget is to compile all estimated costs and allocated resources into one comprehensive document. This should include both direct and indirect expenses associated with the project, as well as estimated resources and their associated costs. Most importantly, the budget summary must be presented clearly for all stakeholders to read easily - project sponsors, executives, team members - so it can be understood by everyone involved (best practices for creating project budgets included).

Best Practices for Creating a Project Budget

Here are some best practices to follow when creating a project budget:

1. Involve all stakeholders in the budgeting process, such as project sponsors, team members and subject matter experts.

2. Draw upon historical data and lessons learned from past projects when making budget estimates.

3. Break down costs and resources into smaller, more manageable components to guarantee accurate estimates.

4. Utilize a project management software tool to create and monitor the budget.

5. Monitor and update the budget throughout the project's lifespan to guarantee it stays on track.

7.6 ALLOCATING RESOURCES

Data migration projects necessitate careful planning and resource allocation to guarantee successful completion within budget and timeline. Allocating resources correctly is essential in meeting project goals and objectives; it involves identifying the right people at the right time with enough money.

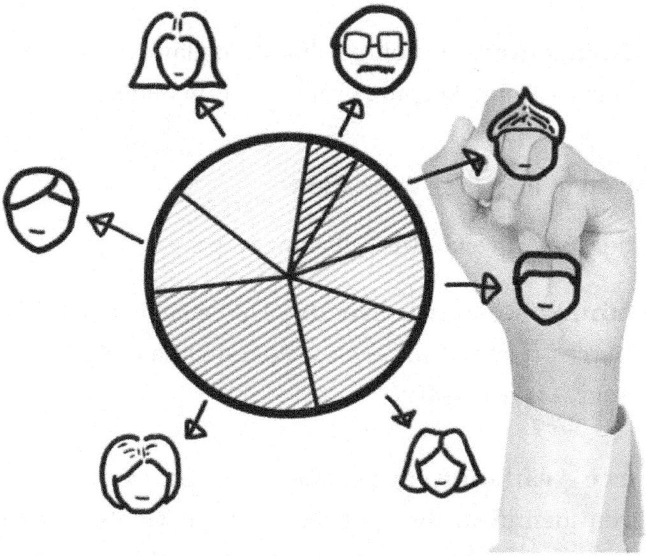

Resource allocation in a data migration project involves assigning staff, equipment and budget to different aspects of the project. It involves making sure each task has sufficient resources to be completed within its scheduled timeline while also keeping within budget.

Resource allocation is an integral aspect of project management and plays a pivotal role in its success. Proper resource allocation is necessary for several reasons:

1. **Maximizing Efficiency**: Resource allocation ensures all resources, such as staff, equipment and budget, are utilized optimally to maximize project completion efficiency.

2. **Achieving Project Objectives**: Allocating sufficient resources to the correct tasks ensures each one is completed effectively and efficiently, ultimately leading to the project's successful conclusion.

3. **Effective Time Management**: Appropriate resource allocation helps manage time efficiently, ensuring tasks are completed within the scheduled timeline.

4. **Cost Management**: Resource allocation plays a significant role in controlling costs and keeping projects within budget.

Steps Involved in Resource Allocation

Resource allocation involves the following steps:

1. **Determining Resource Needs**: This step involves identifying the resources required for each task and estimating how much of each resource is necessary.

2. **Resource Availability**: Once the resources needed for each task have been identified, the next step is to assess their availability. This involves assessing internal resources within the organization or if external assistance will be required.

3. **Allocating Resources:** After resources have been identified and their availability confirmed, the next step is to assign them to specific tasks. Assigning the right resources to the right tasks is paramount for project success.

4. **Monitoring Resource Usage**: It is essential to track resource usage after allocation in order to guarantee they are being utilized efficiently and effectively. This involves tracking staff hours, equipment usage, and budget allocation and making necessary changes as needed.

Challenges in Resource Allocation

Here are some of the challenges in allocating resources:

1. **Limited Resources**: Resource allocation can be problematic due to various reasons, such as limited resources. Organizations may experience difficulties when allocating resources when there are insufficient funds available, leading to conflicts and delays in project completion.

2. **Conflicting Priorities**: Different departments within an organization may have conflicting priorities, leading to conflicts in resource allocation.

3. **Uncertainty**: Lack of certainty regarding resources such as staff or equipment availability can cause delays in project completion.

4. **Shifts in Project Scope**: Shifting the scope of a project may necessitate additional resources or reallocation of existing ones, leading to delays or increased costs.

7.7 PRIORITIZING RESOURCES

Resource allocation is a crucial element in any data migration project. This involves allocating available resources to specific tasks and activities within given constraints such as time, budget, or other limitations. But it's not enough just to allocate resources; prioritize those resources so that critical tasks receive adequate attention and resources.

Prioritizing resources is paramount for any data migration project's success. Here are a few reasons why:

Maximizes Use of Limited Resources:

Data migration projects often face limited resources such as time, budget and manpower. Prioritizing resources ensures they are allocated to the most crucial tasks, maximizing their use and avoiding wastage.

Focuses on Critical Tasks:

Prioritizing resources ensures the most pressing tasks receive adequate attention and resources. This helps guarantee project goals and objectives are achieved within given constraints.

Helps to Identify and Manage Risks:

By prioritizing resources, project managers can identify and address risks associated with their project. Doing this helps guarantee potential issues are addressed before they escalate out of control and negatively affect project outcomes.

Enhances Communication and Collaboration:

Prioritizing resources encourages effective communication and collaboration among team members. By prioritizing tasks, each person can focus on their specific responsibilities while working together towards project success.

Steps for Prioritizing Resources:

Prioritizing resources involves several essential steps. Here are some of the essential ones:

Identify Critical Tasks:

Prioritizing resources begins by identifying the essential tasks for achieving project objectives. These should then be listed in order of priority according to their importance for the project.

Determine the Needed Resources:

Once critical tasks have been identified, the next step is to identify the resources necessary for each one. This includes determining manpower, equipment and materials.

Allocating Resources:

The third step is to allocate resources to the most pressing tasks. This involves allocating available resources according to priority, so that those tasks with the greatest need receive sufficient support.

Monitor and Adjust:

The final step in prioritizing resources is to monitor project progress and adjust resource allocation as necessary. This helps guarantee that the project stays on track, with critical tasks completed within given constraints.

7.8 MONITORING RESOURCE UTILIZATION

Resource utilization is a critical aspect of any project, involving tracking and managing resource use to guarantee they are being utilized efficiently and effectively. In today's world where organizations must do more with less, this concept of resource utilization has become even more essential. In this chapter we will examine why monitoring resource usage is so important and provide some best practices for doing so.

Why Is Monitoring Resource Utilization Essential? Monitoring resource utilization is important for several reasons.

1. **Optimize Resource Allocation**: By monitoring resource utilization, you can identify which resources are being utilized most and underutilized. This knowledge allows for optimized resource allocation so that all resources are being utilized optimally. This ensures you get the most from your resources.

2. **Manage Costs Effectively**: Monitoring resource utilization can help you control costs by recognizing areas where you may be overspending. By recognizing where resources are underutilized, you can reallocate them to areas where they are most needed, thus avoiding unnecessary expenditures.

3. **Define Bottlenecks**: Monitoring resource utilization can help you identify bottlenecks in your project, where resources are being overutilized and resulting in delays. With this data, you can take corrective action to eliminate these obstructions and keep the project moving forward.

4. **Improve Decision Making**: By having an overview of how your resources are being allocated, you can make informed decisions about where to allocate them in the future. Doing this helps you avoid costly errors and guarantee that your resources are being utilized most efficiently possible.

Best Practices for Monitoring Resource Utilization:

Here are some best practices for monitoring resource utilization:

1. **Establish Baseline Metrics**: Before you can begin tracking resource utilization, establish baseline metrics to measure against. Doing this will give your monitoring a starting point and allow you to identify trends over time.

2. **Implement Automated Tools**: Manual monitoring of resource utilization can be time-consuming and error prone. Instead, consider using automated tools that offer real-time data on resource consumption.

3. **Establish Clear Roles and Responsibilities**: Make sure everyone on the project team comprehends their roles and responsibilities when it comes to resource utilization monitoring. Doing this will guarantee everyone works towards achieving similar objectives.

4. **Consistently Review Utilization Data**: Schedule regular reviews of resource utilization data to detect trends and patterns. This will enable you to make informed decisions about resource allocation as well as identify areas for improvement.

5. **Take Action**: Once you've identified resource areas that are underutilized or overutilized, take immediate steps to address them. This could involve reallocating funds, altering processes, or eliminating bottlenecks.

7.9 SUMMARY

- Budgeting and resource allocation refer to identifying the resources necessary for a project and estimating their costs.

- It is essential to determine resources needs by thoroughly examining the scope and objectives of the project.

- Create a resource plan which outlines the types, quantity and timing for all resources required for the project.

- Estimate project costs by accounting for all associated costs related to those resources.
- Establish a project budget which outlines the overall costs and allocates resources accordingly.
- Prioritizing resources is essential to guarantee critical project activities receive sufficient funding.
- Monitoring resource utilization is essential to guarantee resources are used effectively and efficiently throughout the project lifecycle.
- Budget and resource allocation should be reviewed regularly in order to stay on track and make necessary adjustments as needed.
- Effective budgeting and resource allocation can lead to project completion within the anticipated timeline and budget.

7.10 TEST YOUR KNOWLEDGE

I. Which of the following is NOT an important step in identifying resource needs?

a) Assessing project requirements
b) Determining the available budget
c) Identifying team member availability
d) Establishing project goals and objectives

II. Which of the following is the first step in creating a resource plan?

a) Identifying resource needs

b) Allocating resources
c) Monitoring resource utilization
d) Prioritizing resources

III. What is the purpose of creating a project budget?

a) To track project progress
b) To allocate resources
c) To estimate project costs
d) To prioritize resources

IV. Which of the following is an example of a resource constraint?

a) Limited budget
b) Limited team member availability
c) Tight project deadline
d) All of the above

V. What is the primary goal of estimating project costs?

a) To create a budget
b) To allocate resources
c) To identify resource needs
d) To prioritize resources

VI. What is the purpose of prioritizing resources?

a) To allocate resources effectively
b) To track project progress
c) To identify resource needs
d) To estimate project costs

VII. Which of the following is an important consideration when monitoring resource utilization?

a) Team member availability
b) Project requirements
c) Available budget
d) All of the above

VIII. What is the primary purpose of allocating resources?

a) To identify resource needs
b) To estimate project costs
c) To prioritize resources
d) To assign resources to specific tasks

IX. What is the first step in creating a project budget?

a) Identifying resource needs
b) Allocating resources
c) Monitoring resource utilization
d) Estimating project costs

X. What is the primary goal of monitoring resource utilization?

a) To ensure team member availability
b) To track project progress
c) To identify potential resource constraints
d) To prioritize resources

7.11 ANSWERS

I. Answer: d) Establishing project goals and objectives
II. Answer: a) Identifying resource needs
III. Answer: c) To estimate project costs
IV. Answer: d) All of the above
V. Answer: a) To create a budget
VI. Answer: a) To allocate resources effectively
VII. Answer: d) All of the above
VIII. Answer: d) To assign resources to specific tasks
IX. Answer: d) Estimating project costs
X. Answer: c) To identify potential resource constraints

08

8 DATA MAPPING AND EXTRACTION

Welcome to the chapter on "Data Mapping and Extraction". Data migration is an intricate process that involves extracting data from various sources, mapping it onto a target system, then loading it into the new system. The success of any data migration project relies on accurate data mapping and extraction. This chapter will give you a comprehensive understanding of the fundamental concepts, techniques, and best practices related to data mapping and extraction. In this chapter, we'll investigate the challenges associated with data mapping and extraction and how to address them. By the end of it all, you will possess a comprehensive understanding of this process so that you can manage it successfully for a successful data migration project.

8.1 UNDERSTANDING DATA MAPPING

Data mapping is a critical aspect of data migration. It involves mapping data from one system to another, creating a blueprint that ensures accuracy when migrating between systems. Without proper mapping, information may be lost, corrupted or misinterpreted.

Data mapping is essential in data migration projects for several reasons. Here are a few of the top reasons why:

1. **Verifies Accuracy and Completeness of Data**: Data mapping ensures all data from the source system is accurately transferred to

the target system, helping identify missing or incomplete records and guaranteeing all transferred information is correct.

2. **Facilitates Data Transformation**: Data mapping helps transform data from one system to fit into the format and structure of a target system, ensuring that information is correctly interpreted by the new one.

3. **Minimizes Data Errors**: Data mapping helps detect any discrepancies between source and target systems, minimizing errors to guarantee data accuracy during migration. This reduces human errors to a minimum.

4. **Enhances Efficiency**: Accurate data mapping guarantees speedy and effective data migration, saving time and resources while minimizing disruption to business operations.

Steps Involved in Data Mapping

Data mapping comprises several steps which must be completed correctly in order for data migration to take place effectively and quickly.

Data mapping is a multi-step process that involves:

1. **Determining data sources to be migrated**: The initial step in data mapping requires identifying the sources from which data will be extracted. This involves identifying which systems need to be utilized for extraction.

2. **Analyzing Source Data**: The next step in this process is to analyze the source data. This involves comprehending its structure and format within the source system.

3. **Determining Target Data**: The next step requires you to recognize and understand the target data. This involves comprehending its structure and format within the targeted system.

4. **Generating a Mapping Document**: The fourth step requires creating a mapping document, which links data elements from one system to another. This document serves as the connecting link between both systems.

5. **Validating the Mapping**: After testing the mapping, it's time to confirm its accuracy and ensure data migration from source system to target system is successful.

Best Practices for Effective Data Mapping

Testing your mapping ensures accuracy of mapping from source system to target system before proceeding.

To guarantee effective data mapping, the following best practices should be employed:

1. **Recognize business requirements before starting data mapping**: It is essential to comprehend the needs of your business before beginning data mapping. Doing this ensures that the mapping process meets their requirements.

2. **Document the Mapping Process**: Documenting the mapping process is essential for providing a comprehensive understanding of it, making it simpler to resolve any issues that may arise during migration. This also guarantees an efficient workflow throughout all stages of mapping.

3. **Include Stakeholders**: It is critical to involve stakeholders in the mapping process. Doing so ensures their voice is heard during this step, and they have a comprehensive understanding of what needs to be done.

4. **Validate the mapping**: Testing the mapping is essential in guaranteeing that data is accurately migrated from one system to the other.

8.2 TYPES OF DATA MAPPING

Data mapping is an essential part of the data migration process. It involves identifying the data elements in the source system, mapping them to the corresponding data elements in the target system, and defining how the data will be transformed during the migration process. There are different types of data mapping that can be used depending on the complexity of the data migration project. In this section, we will explore the different types of data mapping and their applications in data migration projects.

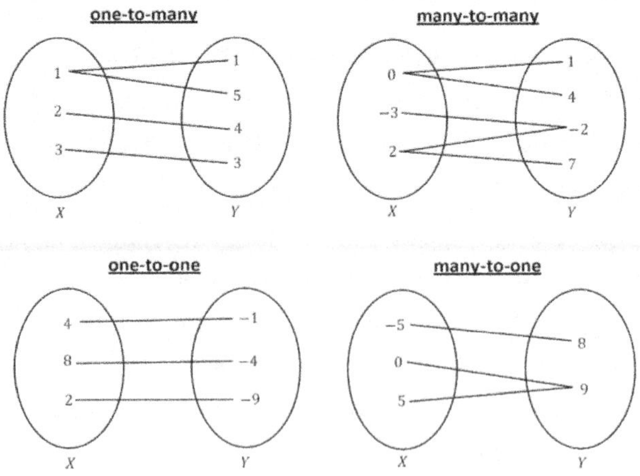

1. **One-to-One Data Mapping**: One-to-one data mapping is the simplest form of data mapping. In this type, a single field in one system is mapped to another identical field in another system without need for data transformation. One-to-one mapping is often employed for straightforward data migration projects with limited complexity.

2. **Many-to-One Data Mapping**: Many-to-one data mapping is employed when multiple elements in the source system are mapped to a single value in the target system. This type of mapping often occurs when source data is more specific than expected and limited fields exist on the target side. Many-to-one mapping has become popularized during data migration projects that require limited fields for conversion purposes.

3. **One-to-Many Data Mapping**: One-to-many data mapping refers to a process in which one data element from the source system is mapped onto multiple elements in the target system. This type of mapping often occurs when source data is less granular than anticipated and more fields are needed in the destination system than what was originally provided. It can also be employed in data migration projects where more fields exist than expected in either direction.

4. **Complex Data Mapping**: Complex data mapping is necessary when the data elements in two systems have distinctly different structures or formats, necessitating transformation during migration. This type of mapping involves complex rules and logic to guarantee correct mapping between systems. It's often employed in projects involving high levels of complexity such as migrating from legacy systems to modern ones.

One real-world application of type-to-type data mapping can be seen in the banking industry, where banks often need to migrate data from one system to another and require mapping data fields between both systems. For instance, a bank may need to transfer customer information from an old account management system to a new one. In the former system, all customer names were stored in one field; however, in the latter one, each name needs to be divided into first and last name fields. In such a scenario, type-to-type data mapping could be employed to ensure accurate migration of customer information from the old system to the new one and that name fields were separated correctly into two distinct fields. Without accurate mapping, banks could risk misplacing important customer details or inputting errors into their new system.

8.3 DATA EXTRACTION TECHNIQUES

Data extraction is the process of taking data from various sources and transforming it into a format suitable for analysis or other uses. Its significance cannot be overstated, as data extraction forms the foundation upon which many analytical and decision-making processes rest. In this section, we'll look at various techniques available to extract information from different sources.

1. **Web Scraping**: Web scraping is a technique used to extract data from websites. It involves using software to grab information from web pages and convert it into structured format. Businesses often employ this technique for gathering competitor information or collecting customer reviews.

2. **Database Extraction**: Database extraction is the practice of extracting data from databases. This can be accomplished either through exporting the information in a structured format or employing tools such as SQL queries. Businesses often utilize this

technique to retrieve information from either their own databases or third-party databases.

3. **ETL (Extract, Transform and Load)**: ETL stands for extract, transform and load and is the process of extracting data from various sources, structuring it in a structured format and then loading it into either a data warehouse or database. This technique is commonly employed in business intelligence projects as well as data warehousing operations.

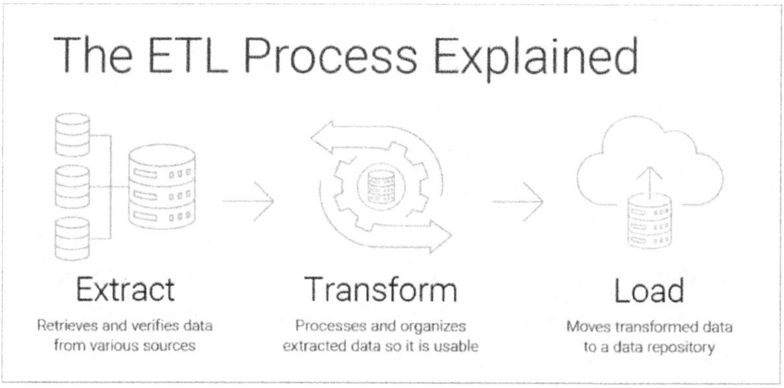

4. **Log File Extraction**: Log file extraction is a technique for extracting data from log files generated by software applications or web servers. It's commonly employed to analyze user behavior on websites or troubleshoot software applications.

5. **Email Extraction**: Email extraction is the process of taking data from emails. This can be accomplished using email clients such as Microsoft Outlook or software tailored specifically for email extraction. Businesses often utilize this technique to collect customer information or analyze consumer behavior.

6. **OCR (Optical Character Recognition):** OCR stands for optical character recognition and it's a technique for extracting data from

scanned documents or images. OCR software can be employed to extract text from documents like invoices or receipts.

7. **Cloud Extraction**: Cloud extraction is the process of taking data from cloud-based applications like Salesforce or Dropbox and extracting it for analysis. This can be done either with APIs or software designed specifically for cloud extraction. Businesses often utilize this technique to collect customer information or analyze sales figures.

8. **Mobile Device Extraction**: Mobile device extraction involves extracting data from mobile devices like smartphones and tablets using software designed specifically for this task. Law enforcement agencies often enlist this technique during investigations to obtain evidence from these mobile phones.

Data extraction is a fundamental step in the data analysis process. Businesses must select an effective technique for extracting information from various sources, and by doing so they will gain valuable insights from their data which in turn allows them to make more informed decisions.

8.4 IMPORTANCE OF DATA MAPPING AND EXTRACTION

Data mapping and extraction are essential steps in data migration projects. The process involves extracting data from one system, mapping it into the format required by the target system, then loading it into that same system. The success of a data migration depends heavily on how accurately and completely this extracted information has been mapped.

In this section, we will address the significance of data mapping and extraction in data migration projects. We'll explain how these processes can influence project outcomes, as well as provide best practices to guarantee accurate extraction and mapping of data.

1. **Verifies data accuracy and completeness**: Data mapping and extraction are essential in guaranteeing the accuracy and completeness of data during migration. Data is structured differently in different systems, so mapping helps understand how to extract and load it into your target system. By mapping data, you can identify any discrepancies or missing information that needs addressing before migration begins.

2. **Reduces Data Loss Risk**: Data mapping and extraction can help minimize data loss during migration by accurately identifying which data needs to be transferred, so all necessary information can be extracted from the source system without missing anything during transfer. This minimizes the chance for data loss during transit.

3. **Enhances Data Quality**: Data mapping and extraction can help enhance data quality. By mapping the data, you can identify any inconsistencies or errors that need to be rectified. Cleaning up the data before migration ensures that the loaded information into the target system is of high quality.

4. **Saves Time and Effort**: Accurate mapping the data can save time during the migration process, helping identify necessary information to be extracted from the source system, saving unnecessary time spent on extracting unnecessary information. Furthermore, mapping reduces errors during transference by increasing precision.

5. **Facilitates Data Integration**: Data mapping and extraction can simplify data integration between systems by identifying how related and how it should be integrated with other data in the target system. Doing this helps guarantee that all data is correctly integrated, with any dependencies resolved before beginning migration processes.

8.5 MAPPING SOURCE & TARGET DATA STRUCTURES

When migrating data, mapping the source and target data structures is a vital step. This involves recognizing the different types of information present in the source system, then figuring out how this should be transformed and mapped into the desired destination system. In this section, we'll look at the various steps involved with mapping sources to target data structures as well as its advantages and drawbacks.

Understanding the Source System Data Structures

A key step in mapping sources to target data structures is gaining an in-depth comprehension of the structures present within the source system. This involves comprehending various entities, their relationships and attributes for each one. Furthermore, any data constraints or business rules that apply to that data must also be taken into account.

Data profiling can provide a deep insight into the source system's data structures, so it may be necessary to conduct data analysis. This involves examining data within the source system in order to spot patterns and relationships as well as any data quality issues or inconsistencies that may exist. Data profiling also assists in detecting potential data mapping problems early on in the project - saving time and resources in the long run.

Mapping Data to the Target System

After an in-depth analysis of source system data structures, the next step is mapping this information onto the desired target system. This involves identifying corresponding data entities within the destination system and then deciding how the information from the source system should be transformed in order to satisfy its demands.

Data mapping can be a complex process, especially if the source and target systems have disparate data structures or models. To make things simpler, using visual mapping software such as MapMyWorld can be beneficial; it helps identify relationships between various entities and attributes.

One of the primary challenges when mapping data to a target system is making sure it's transformed correctly. This involves making sure data is converted into its correct type, any business rules or constraints are applied correctly, and any data transformations occur in their intended order. It's essential to thoroughly test this process in order to guarantee accurate transformations every single time.

Data Mapping and Extraction

Mapping source and target data structures has numerous advantages. One major advantage is that it guarantees accurate migration of data from one system to another, helping minimize any risk of data loss or corruption during migration. Additionally, mapping source and destination systems also helps create more efficient processes by eliminating manual intervention during transfer.

Data mapping and extraction also helps to enhance data quality by detecting and rectifying any quality issues within a source system. By mapping the data to its intended destination, it becomes possible to detect missing or incomplete information as well as inconsistencies or errors in the stored information.

Another advantage of data mapping and extraction is its potential to reduce both cost and duration of a data migration project. By recognizing any potential data mapping issues early in the process, adjustments can be made to the migration plan which helps avoid costly delays or rework later on.

8.6 IDENTIFYING DATA EXTRACTION REQUIREMENTS

Data extraction is a fundamental step in the data migration process. It involves retrieving information from one system and transferring it to another. Data extraction can be laborious, so it's essential that all necessary information is extracted accurately and completely.

The initial step in data extraction is to identify the required extraction requirements. This requires analyzing both the source system and data being extracted. Here are some key elements involved in recognizing data extraction needs:

1. **Analyze the Source System**: Analyzing a source system is the initial step. This involves understanding its structure and layout as well as what data types are stored there. Furthermore, one must comprehend both its data model and how data is stored within it.

2. **Define the Data to be Extracted**: After an analysis of the source system, it is time to identify what data needs extracting. This involves comprehending the business requirements for data migration project and recognizing which data elements are necessary in order to fulfill those demands.

3. **Determining the Extraction Method**: Once you've identified the data to be extracted, the next step is to decide on an extraction technique. There are various options for data extraction such as full data extraction, incremental data extraction and selective data

extraction; select whichever option best meets your data requirements and source system capabilities.

4. **Plan the Data Extraction Process**: Once an extraction method has been chosen, the next step is to plan its execution. This involves creating a data extraction plan that outlines all relevant details such as data elements to be extracted, the extraction technique employed, and any other relevant details.

Best Practices for Determining Data Extraction Requirements:

Here are some best practices to follow when identifying data extraction requirements:

1. **Include Key Stakeholders**: It is critical to include key stakeholders, such as business users, IT staff and data owners, in the data extraction process. Doing this guarantees all relevant requirements are identified and addressed.

2. **Comprehend the business requirements**: It's essential to comprehend the business requirements for a data migration project in order to guarantee that the extraction process meets those needs.

3. **Adopt a structured approach**: Adopting a structured approach when identifying data extraction requirements helps guarantee that all pertinent information is identified and extracted correctly.

4. **Validate the data extraction process**: Before beginning any data extraction project, it's essential to test it thoroughly. This helps identify any problems or errors during the procedure and guarantees that data is extracted accurately and completely.

8.7 PREPARING FOR DATA EXTRACTION

Preparing for data extraction begins with understanding the source systems and databases from which it will be extracted. It is essential to comprehend data structures, schemas, and relationships between tables and fields so that the migration team can accurately identify required information while avoiding unnecessary extraction of irrelevant or duplicate data.

Identifying Data Requirements:

Once the source systems have been identified, the next step should be to pinpoint specific data requirements. The data migration team should collaborate with business stakeholders and subject matter experts to identify this information for the new system - this includes delineating data elements, mapping fields, and creating relationships between them.

Assessing Data Quality:

Data quality is a fundamental aspect of successful data migration. Prior to extraction, it's necessary to assess the quality of the migrated data for any inconsistencies, redundancies or inaccuracies. The data migration team should collaborate with data owners in order to clean and prepare this information for extraction.

Establishing a Data Extraction Plan:

Once the data requirements and quality have been defined, the data migration team can create an extraction plan. This should include the tools/ techniques to be used, an extraction schedule, and data format. Furthermore, it identifies roles and responsibilities for team members involved in this process.

Preparing the Infrastructure:

A robust data extraction infrastructure is necessary to handle large data sets. To make sure all necessary hardware and software resources are available, such as adequate storage capacity, network bandwidth, and processing power are allocated, the data migration team should take measures to guarantee these conditions exist.

Data Extraction Testing:

Prior to beginning the actual extraction process, it is essential to test your plan and procedure. This involves conducting a trial extraction in order to identify any issues or challenges and refine the extraction procedure further. The data migration team should collaborate with business stakeholders and subject matter experts in order to validate extracted data for accuracy and completeness.

8.8 MANAGING DATA MIGRATION SCRIPTS AND CODE

Managing data migration scripts and code is essential for guaranteeing the accuracy and integrity of migrated data. Here are some key reasons why managing these processes is so important:

1. **Maintain Data Integrity**: Data integrity is of the utmost importance during data migration, so it's essential that migration scripts are designed with this in mind. Managing your migration scripts and code ensures they are accurately designed and correctly executed, helping ensure accurate data processing.

2. **Minimize Errors**: Mistakes during data migration can result in data loss or incorrect results, both of which have serious repercussions for organizations. Managing data migration scripts

and code helps minimize errors by ensuring they are designed accurately and executed without errors.

3. **Guarantee successful migration**: Achieving a successful data migration necessitates careful planning, preparation and execution. Managing data migration scripts and code is essential to guaranteeing the process is executed flawlessly.

4. **Reduce Migration Time**: Effective data migration scripts and code management can help minimize migration time by ensuring they are designed and executed efficiently, helping organizations save both time and money during the migration process.

Best Practices for Managing Data Migration Scripts and Code:

Here are some best practices for managing data migration scripts and code:

1. **Utilize a version control system**: Version control systems enable organizations to track changes made to scripts and code over time, making it simpler to track modifications and roll back to earlier versions if needed.

2. **Verify Migration Scripts Thoroughly**: Testing migration scripts thoroughly helps detect errors and guarantees they have been written accurately.

3. **Utilize a data migration tool**: Utilizing an efficient data migration tool can help manage migration scripts and code more efficiently, decreasing errors and guaranteeing a successful migration.

4. **Document Scripts and Code:** Documenting migration scripts and code helps guarantee they are executed correctly, making them understandable to other team members.

Utilizing a data migration tool for managing scripts and code: Utilizing an automated data migration tool makes this much simpler; simply take screenshots of what needs to be done so your work can be verified more accurately.

Benefits of using a data migration tool to manage data migration scripts and code:

1. **Improved efficiency**: Utilizing a data migration tool to manage scripts and code can offer several advantages, such as improved efficiency: Data migration tools automate many aspects of the migration process, making it faster and reducing errors.

2. **Greater Accuracy**: Data migration tools can help guarantee the precision of migration scripts and code, decreasing the risk of data loss or inaccurate information.

3. **Enhancing Collaboration**: Data migration tools enable multiple team members to work on migration scripts and code simultaneously, improving collaboration and productivity.

4. **Enhancing Security**: Data migration tools can provide enhanced protection for migration scripts and code, safeguarding sensitive information from unauthorized access.

8.9 CHALLENGES IN DATA MAPPING & EXTRACTION

Organizations often enlist data mapping and extraction as a key step in their data migration efforts. Unfortunately, this task can often present obstacles that cause delays, raise costs, or even result in project failure. In this chapter we'll address some of the most prevalent difficulties organizations encounter when performing data mapping and extraction.

1. **Acquiring the Source Data**: One of the primary challenges is a lack of comprehension for source data. This can arise when it is poorly documented or dispersed across multiple systems. Without an accurate view of this source information, determining appropriate mapping and extraction strategies becomes challenging.

2. **Incompatible Data Formats**: One major challenge lies in working with incompatible data formats. Different systems store information in various forms, which makes mapping and extracting that data difficult. Moreover, some records may be kept in proprietary formats that require specialized tools to access.

3. **Data Quality Issues**: Data quality issues can pose a significant challenge to data mapping and extraction. Poor data quality can lead to errors and inconsistencies in extracted information due to errors made during data entry, duplicate records, or missing records. These issues could arise as a result of human errors, duplicate entries, or missing records.

4. **Complex Business Rules**: Complicated business rules can make data mapping and extraction more challenging. In certain scenarios, data may need to be transformed based on these complex rules, making the mapping and extraction process even more challenging.

5. **Limited Resources**: Limited resources can also present a challenge when mapping and extracting data, particularly large

datasets. Organizations may need to allocate additional personnel in order to guarantee the process is completed on schedule and within budget.

6. **Evolving Requirements**: Evolving requirements can present a challenge. As business needs alter, data mapping and extraction requirements may also shift. This could cause delays and additional costs as organizations attempt to adjust their strategies for data mapping and extraction.

7. **Testing and Validation**: Testing and validation are essential steps in the data mapping and extraction process, yet they can also be time-consuming and resource intensive. Organizations may need to devise testing and validation strategies in order to guarantee that extracted information is accurate and complete.

8.10 SUMMARY

- Data mapping is the process of connecting data fields from one system to their corresponding fields in another, whilst *Data extraction refers to retrieving information from a source system for migration into an intended target system.
- There are two types of data mapping: manual and automated.
- Data mapping and extraction are critical steps in ensuring accurate and comprehensive data transfer.
- Mapping source and target data structures involves identifying corresponding fields within both systems.
- Determining data extraction requirements involves identifying what needs to be extracted and how.

- Preparing for data extraction involves assuring data quality, selecting an appropriate extraction method, and testing the extraction process.

- Effective management of data migration scripts and code requires version control and testing

- Common challenges in data mapping and extraction involve incompatible data structures, incomplete data sets, and poor-quality data.

8.11 TEST YOUR KNOWLEDGE

I. What is the purpose of data mapping?

a) To identify data extraction requirements
b) To compare source and target data structures
c) To manage data migration scripts and code
d) To identify common challenges in data mapping and extraction

II. What are the two main types of data mapping?

a) Logical and Physical
b) Structured and Unstructured
c) Active and Passive
d) Static and Dynamic

III. Which data extraction technique is suitable for extracting data from large databases?

a) Full extraction
b) Incremental extraction
c) Parallel extraction
d) Multi-table extraction

IV. What is the first step in preparing for data extraction?

a) Identifying data extraction requirements
b) Mapping source and target data structures
c) Creating data extraction scripts
d) Testing the extraction process

V. **What is the most common challenge in data mapping and extraction?**

a) Identifying data extraction requirements
b) Managing data migration scripts and code
c) Ensuring data integrity during extraction
d) Dealing with complex data structures

VI. **Which technique is used for mapping data that is stored in different formats?**

a) Data profiling
b) Data parsing
c) Data scrubbing
d) Data transformation

VII. **Which type of data mapping involves defining relationships between tables?**

a) Logical mapping
b) Physical mapping
c) Structural mapping
d) Hierarchical mapping

VIII. **Which data extraction technique extracts only the data that has changed since the last extraction?**

a) Full extraction
b) Incremental extraction
c) Parallel extraction
d) Multi-table extraction

IX. **Which step in data mapping and extraction involves converting data from one format to another?**

a) Data profiling
b) Data parsing
c) Data scrubbing
d) Data transformation

X. Which challenge in data mapping and extraction involves ensuring that data is extracted correctly?

a) Dealing with complex data structures
b) Managing data migration scripts and code
c) Ensuring data integrity during extraction
d) Identifying data extraction requirements

8.12 ANSWERS

I. Answer: b) To compare source and target data structures
II. Answer: a) Logical and Physical
III. Answer: a) Full extraction
IV. Answer: a) Identifying data extraction requirements
V. Answer: d) Dealing with complex data structures
VI. Answer: d) Data transformation
VII. Answer: a) Logical mapping
VIII. Answer: b) Incremental extraction
IX. Answer: d) Data transformation
X. Answer: c) Ensuring data integrity during extraction

09

9 DATA PROFILING AND CLEANSING

Data is one of the most valuable assets for any organization, yet to maximize its potential, the data must be accurate, consistent and of high quality. Data profiling and cleansing are two essential steps in the data migration process that guarantee this accuracy and consistency.

In this chapter, we'll cover the importance of data profiling and cleansing in data migration. We'll also introduce you to basic concepts and techniques related to data profiling and cleansing, including how to detect and address data quality issues. Furthermore, we'll look at various tools used for profiling and cleansing as well as provide practical tips on achieving successful data migration through effective profiling and cleansing. By the end of it all you'll have a solid grasp on how data profiling and cleansing can improve your data's quality as well as contribute towards project success by effectively improving its efficiency.

9.1 INTRODUCTION TO DATA PROFILING

Data is being generated at an unprecedented pace, reinforcing the significance of data profiling. Data profiling is the process of analyzing, reviewing and understanding data to guarantee its quality, completeness and accuracy. It plays a significant role in data migration/integration projects as well as ongoing data management initiatives.

At its core, data profiling involves analyzing the characteristics of data to uncover its structure, patterns and relationships. This includes recognizing data types, constraints and connections between fields and tables as well as checking for errors, inconsistencies or redundancies within it.

One of the key advantages of data profiling is its ability to detect potential data quality issues early in their lifecycle, before they become bigger issues. By analyzing and understanding data, organizations can detect errors or anomalies and take appropriate actions for correction. This guarantees accurate and dependable information which in turn allows for improved decision-making and more efficient business processes.

Data profiling can also be employed to guarantee compliance with regulatory obligations such as the General Data Protection Regulation (GDPR) or Health Insurance Portability and Accountability Act (HIPAA). By understanding data, organizations can identify sensitive information that should be protected and implement suitable data security measures.

Another application of data profiling is in data integration projects. When migrating data from one system to another, it's essential that it remains compatible with the new one. Data profiling can help detect any incompatibilities or quality issues during migration.

Overall, data profiling is a crucial process for any organization that strives to guarantee the quality, completeness, and accuracy of its data. By analyzing and understanding this information, companies can detect potential issues early and take steps to rectify them; this leads to improved decision-making processes, more efficient business operations, as well as higher compliance with regulatory requirements.

9.2 BENEFITS OF DATA PROFILING

Data profiling is an integral component of data migration. It involves analyzing, understanding and assessing data from various sources to provide valuable insights into its quality, structure and contents - essential for any successful data migration project. In this section we'll explore the advantages of data profiling and how it can assist organizations with their initiatives.

Benefits of Data Profiling:

Data Profiling Offers the Following Advantages:

1. **Recognizing Data Quality Issues**: Data profiling can assist in recognizing data quality problems such as incomplete or missing data, inconsistent records, and duplicate records early in the migration process. By recognizing these problems early on in the migration effort, organizations save both time and money in the long run.

2. **Better Data Understanding**: Data profiling provides a deeper insight into data structure, contents and relationships by analyzing its attributes. This allows organizations to make educated decisions regarding how best to manage, cleanse and transform their records.

3. **Improved Data Governance**: Data profiling can assist organizations in creating and adhering to better data governance practices. It allows them to detect and address data quality issues - an essential aspect of effective data governance.

4. **Improved Data Integration**: Data profiling can assist organizations in more efficiently integrating data from various sources. By recognizing data quality issues, data profiling allows organizations to guarantee that the combined information from different sources is accurate and efficient.

5. **Risk Mitigation**: Data profiling can assist organizations in mitigating the risks associated with data migration. By recognizing

and correcting data quality issues early on in the process, organizations can reduce the chance of data loss or corruption during migration.

6. **Improved Decision Making**: Data profiling provides organizations with valuable insights into their data that can assist in making more informed decisions. By comprehending the data more thoroughly, organizations are able to make decisions based on accurate and dependable information.

7. **Time and Cost Savings**: Data profiling can help organizations save both time and money in the long run by detecting data quality issues early on. Addressing these problems early on helps organizations avoid costly, time-consuming data remediation initiatives down the line.

Use Case:

A healthcare organization was facing difficulties with their data migration project. This involved moving data from multiple systems onto a new platform, but the organization had difficulty understanding the data coming from various sources and ran into data quality issues which caused delays in progress.

To address these challenges, the organization implemented data profiling. Data profiling tools were utilized to analyze and comprehend information from various sources, enabling them to detect data quality issues such as missing or incomplete data and inconsistent records early in the migration process. Doing this saved them costly and time-consuming remediation initiatives later on.

Data profiling provided invaluable insights into the data, enabling the healthcare organization to make more informed decisions about how to manage, cleanse and transform it. By employing data profiling, they were able to complete their data migration project on schedule with improved data quality at no additional cost.

9.3 TECHNIQUES FOR DATA PROFILING

Data profiling is an essential step in the data migration process. It analyzes and understands the data structure, content, and quality so as to identify issues with it before migration begins. Data profiling also assists with recognizing patterns, relationships, and inconsistencies within it. In this chapter we'll cover different techniques used for data profiling:

Overview of Data Profiling Techniques:

Data profiling techniques can be broadly divided into two categories:

Statistical Techniques:

Statistical techniques involve analyzing data with mathematical and statistical methods to detect patterns, relationships, and discrepancies. These algorithms calculate frequency distributions, means, medians, modes, and standard deviations from the data set. The results from these analyses help detect anomalies or discrepancies within it. Some commonly employed statistical techniques include:

- **Descriptive Statistics**: This technique involves summarizing data using statistical measures like mean, median, mode and standard deviation.

- **Inferential Statistics**: This involves making predictions about the population based on sample data.

- **Correlation Analysis**: This technique involves examining the correlation between two variables in order to uncover their connection.

- **Regression Analysis**: Regression analysis investigates the relationship between a dependent variable and one or more independent variables by analyzing its relationship with a dependent variable.

- **Time Series Analysis**: This technique involves examining data over time in order to detect trends and patterns.

Data Mining Techniques:

Data mining techniques utilize machine learning algorithms to detect patterns, relationships and anomalies in data. They employ techniques like decision trees, clustering, association rules and neural networks as a way of detecting inconsistencies and anomalies within the dataset. Some common examples of data mining techniques include:

- **Decision trees**: This technique involves creating a tree-like structure to represent decisions made based on input data.

- **Clustering**: Clustering data into clusters based on similarities between them helps with decision making.

- **Association Rules**: This technique involves recognizing the connections between various variables in data.

- **Neural Networks**: Utilizing neural networks, one can construct a network of nodes that can learn from experience and make predictions based on that learning.

9.4 DATA PROFILING TOOLS AND SOFTWARE

In today's digital age, businesses generate an abundance of data which can come from various sources and formats. Before this data can be utilized for analysis, it must first be processed through steps such as data profiling and cleansing. Data profiling refers to examining datasets to gain a comprehensive understanding of their characteristics and guarantee its quality; it's crucial for companies to guarantee their data is accurate, comprehensive, and consistent.

Data profiling can be done manually, but the process can be time-consuming and prone to errors. Fortunately, several software solutions exist that automate this task and make it more efficient. In this section we'll review some of the popular data profiling tools and software options currently available on the market.

Popular Data Profiling Tools and Software:

1. **IBM InfoSphere Information Analyzer**: This powerful tool conducts an in-depth analysis of the quality of data by detecting anomalies, recognizing patterns, and verifying compliance with data quality standards. Furthermore, users have the option to customize the profiling process according to their individual requirements.

2. **Talend Data Quality**: This open-source data quality tool offers features like data profiling, cleansing and enrichment. Additionally, it can perform address validation, deduplication and data masking operations.

3. **Informatica Data Quality**: This software offers a suite of data quality applications to assist with data profiling, cleansing and governance. Furthermore, its metadata-driven approach simplifies the creation and management of data quality rules.

4. **Trifacta**: This cloud-based data management solution offers a comprehensive data profiling solution, including data discovery, cleaning and transformation. It utilizes machine learning

algorithms to automatically recognize patterns and anomalies within data.

5. **Oracle Enterprise Data Quality**: This solution offers a comprehensive suite of data profiling and quality management features to help organizations guarantee the accuracy and completeness of their data. It also contains data governance tools for managing data lineage, cataloguing, and stewardship.

Use Case:

A major e-commerce company was experiencing several issues related to their customer data. The information was incomplete, inconsistent and contained duplicates which resulted in poor customer experience and lost revenue. To rectify the situation, they implemented Talend Data Quality's data profiling tool which allowed them to analyze customer records thoroughly and identify the root causes of poor quality. Through cleansing and enriching existing records with this new software, companies were able to create more accurate customer databases; ultimately leading to improved service, increased loyalty among customers, and higher revenues overall.

9.5 UNDERSTANDING DATA QUALITY

Data quality refers to the accuracy, completeness, consistency, relevance and timeliness of data. In other words, it measures whether or not an object can be trusted for its intended purpose. Data quality isn't a one-off activity but an ongoing process that necessitates continuous monitoring and improvement.

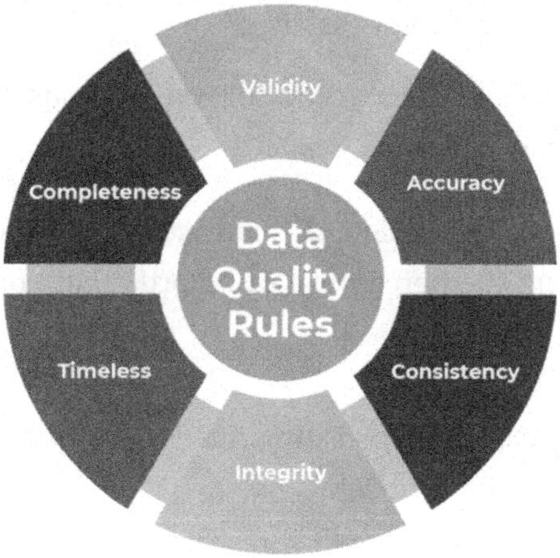

Factors Affecting Data Quality:

Data quality can be affected by a number of factors. Some major ones include:

1. **Data entry errors**: Data entry errors occur when incorrect information is entered into the system due to human or system mistakes or other external causes.

2. **Incomplete Data**: Incomplete data refers to information that is missing due to errors in data entry, system limitations or other external factors. This can occur due to human or machine errors as well as technical malfunctions.

3. **Inconsistent Data**: Data that is inconsistent or irregular can arise when different individuals enter it differently or the data isn't standardized across an organization. This type of issue arises when people enter different information into different systems.

4. **Duplicate Data**: Duplicate data refers to information that exists in multiple locations or systems. This can occur when data is entered multiple times or copied from one system to another.

Methods for Measuring Data Quality:

Data quality can be evaluated using several methods. Some of the more widely-used metrics include:

1. **Completeness**: Completeness measures how well data is filled out, measuring the percentage of records that contain all required information.

2. **Accuracy**: Accuracy measures the degree to which data is accurate, or free from errors.

3. **Consistency**: Consistency assesses how well a given piece of information remains consistent over time, measuring uniformity and standardized values.

4. **Timeliness**: Timeliness measures the degree to which data is up-to-date and relevant.

Best Practices for Ensuring Data Quality:

To guarantee data quality, it's essential to adhere to industry best practices. Some of these best practices for guaranteeing data quality include:

1. **Establish Data Quality Standards**: Establish data quality standards that specify acceptable levels of data quality so all employees in your organization adhere to the same expectations and practices. This will guarantee consistent outcomes from every investigation undertaken.

2. **Conduct regular data audits**: Conducting regular data audits helps identify data quality issues and guarantees the data is accurate, complete, and consistent.

3. **Implement data validation checks**: Utilize data validation checks to detect and rectify errors before they have a detrimental impact on the organization.

4. **Use data profiling tools**: Data profiling tools can be employed to assess and detect data quality issues. These instruments may assist in recognizing data errors, inconsistencies, and redundancies.

9.6 METHODS FOR MEASURING DATA QUALITY

Data is one of the most invaluable assets for organizations in today's digital age. Organizations use it to gain insights, make decisions and manage business operations more efficiently. Unfortunately, inaccurate or incomplete data can lead to inaccurate analysis and decisions which could result in major financial and reputational losses. Thus, it's critical that data quality be guaranteed prior to using it for analysis or decision-making purposes.

Data quality is the measure of data accuracy, completeness, consistency, validity and timeliness. In this chapter we will examine various methods for assessing data quality.

Data Profiling:

Data profiling is the process of examining data to assess its quality and structure. It's an essential step in measuring data quality as it provides an overview of any issues present. Profiling helps detect inconsistencies, missing data, duplicate records, as well as other quality concerns which could affect accuracy or completeness of the collected information.

Data Cleansing:

Data cleansing is the process of correcting or erasing inaccurate, incomplete, and inconsistent data. It plays a significant role in measuring data quality by improving its accuracy and completeness. Data cleansing typically entails various techniques like standardization, de-duplication, and enrichment.

Data Validation:

Data validation is the process of verifying that data meets required standards and specifications. It plays a significant role in measuring data quality, as it guarantees its accuracy, consistency, and completeness. Validating data involves various techniques such as field validation, cross-field validation, and business rule validation.

Data Accuracy:

Data accuracy is a vital aspect of data quality. It indicates the correctness of individual values within a dataset. Data accuracy can be assessed through comparison to trusted sources or by performing statistical analysis to detect inconsistencies.

Data Completeness:

Data completeness refers to the extent to which all required data is present. It plays an essential role in data quality as incomplete records may lead to incorrect analysis and decision-making. Data completeness can be measured by comparing it with a model or performing statistical analysis to detect missing values.

Data Consistency:

Data consistency refers to the degree of consistency among and within data sets. Inconsistent information can lead to incorrect analysis and decision-making, so it's important to measure data consistency using either a data model or statistical analysis in order to detect discrepancies.

Data Timeliness:

Data timeliness refers to the accuracy and relevance of a piece of data. Timing plays an integral role in data quality, especially real-time systems where decisions must be made quickly. You can measure data timeliness by comparing it with a data model or performing statistical analysis to detect outdated records.

Use Case:

A multinational corporation with operations across various countries was facing data quality issues that were impairing its decision-making processes. They were using data from multiple sources, including external ones, but it was inconsistent and incomplete. Without trusting this information, they made decisions based on incomplete or inaccurate details.

9.7 IMPORTANCE OF DATA CLEANSING

Data is an invaluable asset for organizations, yet its quality can often be compromised due to human errors, system malfunctions or data inconsistencies. Data cleansing - also referred to as data scrubbing - is the process of detecting and correcting these errors or inconsistencies in data so that all information is accurate, complete and consistent. In this section we'll examine why data cleansing is so important when migrating data and how it helps organizations reach their business objectives.

Ensuring Accurate Data:

Data cleansing is essential for ensuring accuracy in data. Without accurate information, business decisions cannot be made and performance can suffer significantly. Data cleansing helps detect errors such as misspellings, duplicates, and formatting problems which lead to inaccurate information. With accurate data cleansing services in place, organizations are better able to detect and rectify these issues before they have an adverse impact on their operations.

Improving Data Quality:

Data cleansing is essential for improving data quality. Data quality measures how well a given piece of information meets its intended use, leading to issues like poor decision-making, lost revenue and legal or regulatory violations. Data cleansing helps address these issues by eliminating inconsistencies and guaranteeing the data is complete, accurate and consistent.

Saving Time and Money:

Data cleansing can save organizations both time and money. Inaccurate data can lead to wasted effort trying to fix errors, while poor data quality could mean lost revenue or increased costs. Data cleansing helps eliminate these problems by guaranteeing accurate, high quality data - leading to cost savings as well as improved efficiency.

Preventing Data Breaches:

Data cleansing should also focus on data security. Data breaches can have devastating results for organizations, including financial loss, damage to reputation and legal liability. Data cleansing helps prevent breaches by identifying sensitive data points and removing it securely, thus keeping it under wraps and protected.

Use Case:

In the healthcare industry, data cleansing plays a vital role. Medical records contain sensitive patient information that must remain accurate and up-to-date to guarantee patients receive appropriate treatments. Data cleansing helps identify and fix errors such as misspelled patient names or incorrect medication dosages in medical records - mistakes which could have serious repercussions for patient safety if not rectified.

9.8 DEVELOPING DATA CLEANSING PROCEDURES

Data cleansing is the process of detecting and correcting corrupt or inaccurate records in a dataset. It involves several activities designed to improve quality by eliminating duplicates, filling in missing values, correcting erroneous data, and standardizing format. Developing data cleansing procedures is an integral part of data migration and integration projects as it guarantees reliable accuracy across all systems. In this section we'll outline steps involved with developing these procedures as well as best practices organizations can follow to guarantee successful completion of their cleansing projects.

Establishing Data Cleansing Procedures:

Step 1: Define Data Quality Requirements

The initial step in developing data cleansing procedures is to define what quality standards and metrics must be met for it to meet these expectations. This involves specifying accuracy, completeness, consistency, timeliness as well as other relevant quality criteria that need to be fulfilled during processing. Furthermore, including stakeholders in this process helps guarantee their needs are fulfilled and expectations exceeded.

Step 2: Recognize and Address Data Quality Issues

The next step in data quality assurance is to identify and address any problems. This involves analyzing the data to detect errors, inconsistencies and other concerns that affect its reliability. Common quality issues include missing values, invalid formats, duplicates and discrepancies between sources.

Step 3: Establish Data Cleansing Procedures

Once data quality issues have been identified, the next step is to devise data cleansing procedures to address them. This involves specifying rules and processes for detecting errors, correcting them, filling in missing values, and standardizing data format. It's essential that these protocols be documented clearly and consistently across all systems.

Step 4: Implement Data Cleansing Procedures

Once data cleansing procedures have been designed, the next step is to implement them. This involves applying rules and processes to correct errors, fill in missing values, and standardize formats. It's essential to test these procedures thoroughly to guarantee they work effectively at improving data quality.

Step 5: Monitor and Maintain Data Quality

The final step in data quality management is to continuously monitor and maintain it. This involves creating an ongoing monitoring process, as well as recognizing and addressing new data quality issues as they emerge. Having a governance framework in place is essential to guarantee data quality standards are upheld, along with updating cleansing procedures when needed.

Best Practices for Crafting Data Cleaning Procedures:

1. Involve stakeholders in developing data quality requirements so their needs are met.

2. Utilize a standardized process for recognizing and correcting quality issues.

3. Create data cleansing procedures that are well-documented and simple to follow.

4. Test these cleansing processes thoroughly prior to implementing them.

5. Establish an ongoing system for data quality monitoring and maintenance.

6. Employ data profiling tools to automate the process of identifying data quality issues.

7. Assure that all data cleansing procedures adhere to industry standards and best practices.

Use Case:

A healthcare organization is migrating its patient data from an old legacy system to a new system. As they do so, they discover several data quality issues in the patient records such as missing values, duplicates and inconsistencies between sources. They want to address these issues immediately. To guarantee accurate and dependable data in a new system, an organization creates data cleansing procedures. These steps involve recognizing and rectifying any quality issues such as filling in missing values or eliminating duplicates. The organization tests procedures thoroughly and implements them to resolve data quality issues. Furthermore, it sets up a process for ongoing data quality monitoring and maintenance to guarantee that the data remains accurate and dependable over time.

9.9 HANDLING DATA CLEANSING ERRORS

Data cleansing is a vital element of data migration, where data is transformed and cleaned up to ensure its accuracy and consistency. This step includes identifying and rectifying any inaccuracies in the migrated data to ensure its reliability and accuracy. However, errors may still arise during data cleansing so procedures must be put in place to address them efficiently.

Common Data Cleansing Errors:

This section addresses common errors encountered during data migration and how best to address them.

1. **Missing or Incorrect Data**: Its One of the most frequently occurring errors when cleaning data is missing or incorrect data. This problem occurs when data from source systems or incorrectly entered into migration systems is missing or entered incorrectly, necessitating identification and resolution so as to ensure complete and accurate migration data sets.

2. **Duplicate Data**: Duplicate data can be an easy source of error during data migration processes. Duplicate records must be identified and resolved quickly as they can compromise quality and accuracy issues, creating significant headaches for data teams.

3. **Inconsistent Data**: Inconsistent data refers to information that differs in structure or format from that which was intended, often due to incorrect entry or multiple structures within its source system. Any inconsistencies must be identified and addressed quickly so as to guarantee accurate, uniform migration of migrated data.

4. **Invalid Data**: Invalid data refers to information that does not comply with predetermined data standards and should be identified and addressed immediately as it can lead to issues with quality and accuracy of data. It must also be corrected.

Handling Data Cleansing Errors:

1. **Create Data Cleansing Procedures**: Deliberating on data cleansing procedures is an essential component of handling data cleansing errors effectively. Such steps should outline steps taken to identify and resolve data errors.

2. **Establish Error Reporting Mechanisms**: Establishing error reporting mechanisms is vital to effectively handling data cleansing errors. These mechanisms should enable users to report errors while providing data cleansing teams with the means of tracking and rectifying these problems.

3. **Delegate Responsibilities**: Assigning individual responsibilities is crucial when handling data cleansing errors. They should be responsible for recognizing, correcting and documenting these errors.

4. **Establish Error Resolution Priorities**: Achieving data cleansing errors requires setting priorities based on their impact to both quality and accuracy of data. Priorities can be created based on impact assessments relating to errors that arise during data cleansing activities.

5. **Implement Error Resolution Processes**: Error resolution processes are an indispensable asset when handling data cleansing errors. They should outline all steps required to identify, address, and document data cleansing errors.

9.10 BEST PRACTICES FOR DATA CLEANSING

Data cleansing is an integral component of data migration processes. It involves detecting and correcting inconsistencies from data sets in order to enhance data quality, such as correct business decisions, financial losses, regulatory violations and reputational harm caused by poor quality data. Poor data quality can have severe repercussions such as incorrect business decisions, financial losses, regulatory violations and reputational damage; so adopting best practices for data cleansing to ensure data are accurate, complete, and consistent is paramount for its migration process. In this chapter we'll go over those best practices so they can be implemented during migration processes.

Establish Data Quality Rules:

The initial step of data cleansing is establishing data quality rules. These rules outline what standards must be met for data to be considered clean, such as being complete, accurate, consistent, valid and timely. Once established, these rules can help identify data quality issues and prioritize data cleansing activities.

Use Data Profiling Techniques:

Data profiling can help organizations detect data quality issues and understand their nature and scope of the issue. Data profiling entails analyzing content, structure and relationships of data for patterns, anomalies or inconsistencies which can then be used to develop quality rules or prioritize data cleaning activities.

Develop Cleansing Procedures:

It is necessary to develop data cleansing procedures based on quality rules and results of data profiling. Cleansing can include correction, standardization, matching and enrichment. In order to ensure effective data cleansing it is crucial that such procedures are repeatable, scalable and auditable so as to guarantee consistent and efficient treatment of the data.

Implement Data Cleansing Tools:

Data cleansing tools can automate data cleaning procedures and increase their efficiency and effectiveness, streamlining procedures while also streamlining workflow. Common examples of such tools are data quality software, profiling software, matching software, enrichment software or enrichment modules - it is important to select ones which suit the organization and integrate seamlessly with any data management systems in place.

Monitor Data Quality:

Data quality monitoring is an ongoing process of measuring and reporting on the quality of data over time. It is crucial that this occur to ensure data remains accurate, complete, and consistent after cleansing has occurred. A variety of quality metrics can be defined including number of data quality issues resolved as well as impact of issues on business processes.

Establish Data Governance:

Data governance is a framework for overseeing data quality, privacy, security and compliance. It involves setting policies, standards and procedures related to managing data as well as assigning responsibilities for its administration and monitoring compliance with policies and standards set for data management. Implementing such governance can ensure that data cleansing occurs consistently and transparently, and that quality remains over time.

Training and Support:

Both training and support are integral components of effective data cleansing. Procedures and tools used for cleansing should be documented and made accessible to users; then training should be provided so they understand how to utilize these processes and tools; furthermore, support should also be offered if any issues arise during data cleansing.

9.11 CHALLENGES IN DATA PROFILING & CLEANSING

Profiling and cleansing are integral parts of data migration. Profiling refers to the examination of existing sources for data while cleansing is the act of correcting or eliminating errors and inconsistencies from that source data. Data quality has an immense bearing on migration projects' success - although data profiling and cleansing do not come without their own set of challenges.

Challenges in Data Profiling:

1. **Data Volume**: In today's digital world, organizations generate and store vast quantities of information that is difficult to analyze or profile effectively due to its sheer magnitude. Profiling large amounts of data requires time-intensive effort as well as more resources than expected.

2. **Data Complexity**: Data is often complex in nature, with multiple formats, sources and structures that make analyzing it challenging and its quality compromised if mishandled improperly.

3. **Data Privacy**: Organizations recognize that data privacy is a primary consideration, with sensitive personal and financial data needing to be protected at all costs. They must abide by data protection laws and regulations while profiling and cleansing data sets.

4. **Data Governance**: Data governance refers to the practice of overseeing the availability, usability, integrity, and security of an organization's data. Ineffective governance can lead to challenges with data profiling such as incomplete or inaccurate records being stored within an organization's systems.

Challenges in Data Cleansing:

1. **Data Inconsistencies**: Data inconsistencies can arise due to data entry mistakes, system glitches, or integration issues and can result in incorrect or incomplete data that makes profiling and cleansing it challenging.

2. **Data Duplication**: Data duplication occurs when duplicate entries of identical information are made into multiple systems or taken from multiple sources, making it hard to profile and cleanse in order to improve data quality. This problem poses considerable difficulties for data analysts.

3. **Data Integration**: Data integration refers to the practice of merging information from multiple sources into a centralized view, but can present numerous challenges, including data conflicts, inconsistencies and duplication.

4. **Data Quality Standards**: Settling on proper data quality standards is vital to efficient profiling and cleansing operations, since failure to do so could result in incorrect or missing cleansing

or data being left out altogether, leading to subpar data quality overall.

Best Practices to Address Data Profiling and Cleansing Challenges:

1. **Begin Early**: To successfully overcome data profiling and cleansing challenges, organizations should begin their data profiling early in a migration project so there is enough time for analysis and correction of any potential issues that arise.

2. **Standardize Data**: By standardizing data formats, sources, and structures, standardizing can make profiling and cleansing data simpler - something that data governance processes can facilitate.

3. **Leverage Automated Tools**: Automated tools can assist in speeding up data profiling and cleansing processes, decreasing human errors and increasing accuracy.

4. **Establish Data Quality Standards**: Organizations should establish data quality standards to outline requirements and establish policies and guidelines for data profiling and cleansing.

5. **Perform Regular Data Audits**: Conducting data audits at regular intervals can assist organizations with quickly detecting any inconsistencies and duplication in their data, so corrective action can be taken swiftly to correct it.

9.12 FUTURE OF DATA PROFILING AND CLEANSING

Information has become the new oil, with organizations across industries investing heavily in data collection, analysis and utilization. But as data volumes and complexity increase, so too do challenges associated with managing and assuring its quality - that's where data profiling and cleansing come in - these techniques help organizations identify, analyze and correct any problems with their data to improve its overall quality and increase reliability for use in decision making processes.

As organizations increasingly rely on data-driven insights for business strategy and decision making, data profiling and cleansing will only become increasingly important. In this chapter, we will take a deeper look into its future by considering emerging trends, new technologies, and how this technique will play a part in shaping tomorrow.

1. **Machine Learning and AI**: Machine learning and artificial intelligence have become one of the key trends in data profiling and cleansing, providing rapid analysis of vast volumes of data quickly and precisely, which proves invaluable when it comes to quickly detecting quality issues and developing effective cleansing procedures. As these technologies continue to advance, more sophisticated tools and techniques may appear that further increase efficiency and effectiveness when it comes to profiling and cleansing data.

2. **Automation**: One trend set to change the future of data profiling and cleansing is automation. With ever-increasing volumes and complexity of data, manual processes simply cannot keep pace. Automation helps streamline this process while decreasing time and resources used in identifying and correcting quality issues with data quality issues. We should anticipate seeing more organizations adopt automated profiling and cleansing processes in coming years - making these techniques more accessible.

3. **Integrating With Data Governance**: Data profiling and cleansing aren't standalone processes; rather they're part of an overarching data management and governance framework. Over time, profiling and cleansing should become even more tightly

integrated with governance practices to identify quality issues early in their data lifecycle - this integration should lead to improved overall data quality as well as more reliable decision-making by an organization.

4. **Emphasis on Data Privacy and Security**: As data becomes ever more valuable, its exposure to cyber-attacks and security risks increases significantly. We can expect a stronger focus on data privacy and security when profiling and cleansing, which means implementing more robust protection measures like encryption, access controls, and monitoring to ensure sensitive information always remains protected.

5. **Ongoing development and proliferation of data sources and formats**: In the coming years, we can expect the evolution and proliferation of data sources and formats to accelerate. This may present new challenges for data profiling and cleansing as organizations try to manage and analyze information coming from a wide range of sources such as social media platforms, IoT devices and emerging technologies; but with proper tools and techniques at their disposal they can overcome such challenges to extract value from their data.

9.13 SUMMARY

- Data profiling involves examining data to gain an understanding of its structure, content, and quality.

- Data cleansing refers to the process of detecting and correcting or erasing corrupt, inaccurate, or incomplete data sets.

- Profiling and cleansing data during a migration process are critical steps for ensuring accuracy and completeness of data. Proxying can identify data quality issues, identify patterns in relationships within data, as well as highlight any integration challenges which might be present.

- Methods of measuring data quality include completeness, accuracy, consistency, and conformity.

- Data cleansing procedures involve identifying and resolving data quality issues while creating effective cleansing processes and handling any errors during their implementation.

- For effective data cleansing, best practices involve identifying and prioritizing data quality issues, setting clear data quality goals, and devising an extensive cleansing plan.

- Data profiling and cleansing often present several unique challenges, including inconsistencies among data sets, privacy/security concerns, as well as maintaining data quality over time.

- The future of data profiling and cleansing involves advances in machine learning and artificial intelligence, automated data profiling processes, as well as increased emphasis on data privacy and security.

9.14 TEST YOUR KNOWLEDGE

I. Which of the following best describes data profiling?

a) A process of data analysis that helps to identify patterns and inconsistencies in data
b) A process of extracting data from one source to another
c) A process of cleaning data using advanced algorithms

II. What is the purpose of data cleansing?

a) To extract data from various sources

b) To analyze data and identify inconsistencies
c) To correct or remove inconsistent or inaccurate data

III. What is the main benefit of data profiling?

a) Identifying patterns and inconsistencies in data
b) Extracting data from various sources
c) Cleaning data using advanced algorithms

IV. What is the main challenge in data cleansing?

a) Identifying data quality issues
b) Developing data cleansing procedures
c) Handling data cleansing errors

V. Which of the following is not a technique for data profiling?

a) Data visualization
b) Data aggregation
c) Data cleansing

VI. Which of the following is a data profiling tool?

a) Excel
b) Python
c) Talend

VII. What is the main goal of data quality measurement?

a) To identify patterns and inconsistencies in data
b) To extract data from various sources
c) To measure the accuracy and completeness of data

VIII. Which of the following is a common challenge in data cleansing?

a) Identifying data quality issues
b) Developing data cleansing procedures

c) Handling data cleansing errors

IX. What is the main benefit of using automated data cleansing?

a) Increased accuracy and consistency of data
b) Faster data extraction
c) Improved data visualization

X. What is the future of data profiling and cleansing?

a) It will become less important as data becomes more accurate
b) It will become more important as data volumes continue to grow
c) It will be replaced by advanced machine learning algorithms

9.15 ANSWERS

I. Answer: a) A process of data analysis that helps to identify patterns and inconsistencies in data
II. Answer: c) To correct or remove inconsistent or inaccurate data
III. Answer: a) Identifying patterns and inconsistencies in data
IV. Answer: c) Handling data cleansing errors
V. Answer: c) Data cleansing
VI. Answer: c) Talend
VII. Answer: c) To measure the accuracy and completeness of data
VIII. Answer: a) Identifying data quality issues
IX. Answer: a) Increased accuracy and consistency of data
X. Answer: b) It will become more important as data volumes continue to grow

10

10 RECOMMENDATIONS FOR PRE-MIGRATION ACTIVITIES

Data migration is an intricate process that requires meticulous planning and execution in order to successfully move data from one system to the other. Pre-migration activities play an essential role in guaranteeing its success. This chapter provides organizations with recommendations for pre-migration activities they can undertake to ensure an efficient data migration. Recommendations include creating a comprehensive plan, designing an effective data backup and recovery strategy, adopting industry standards and frameworks, using automation and tools, creating a pre-migration center of excellence, assuring data security and privacy compliance, meeting legal/regulatory compliance requirements, and adhering to legal/regulatory compliance obligations - these will all help minimize risks, lower costs and ensure a successful data migration project.

10.1 DEVELOPING A COMPREHENSIVE PLAN

Planning is crucial to the success of data migration processes. Without careful preparation, an unorganized migration could result in data loss, security breaches and other issues which threaten business operations. Below are a few reasons why creating an exhaustive plan is vital:

1. **Ensuring Data Accuracy**: With an exhaustive plan in place, all data can be identified, analyzed, and verified as being accurate and complete prior to and following migration.

2. **Mitigating Risks**: An exhaustive plan helps identify and reduce potential risks such as data loss, security breaches and system downtime that might arise during or after migration.

3. **Simplify the Process**: An efficient data migration plan will ensure a streamlined, streamlined, and effective data migration process with reduced downtime and costs.

4. **Maintain Schedules**: A comprehensive plan provides a schedule for migration activities, detailing crucial milestones and deadlines so that migration occurs on time.

How to develop a comprehensive plan?

1. **Define Your Scope**: Determine which systems, applications and data need to be migrated before considering its impact on business as well as any regulatory or compliance obligations.

2. **Assess Data Migration Needs**: Evaluate all aspects of data to be migrated, such as quality, completeness and accuracy. Additionally, identify any dependencies, mappings or transformations necessary in the migration.

3. **Determine Your Migration Strategy**: It is essential to select an effective migration strategy which includes details regarding its order, method and tools/technologies necessary for implementation.

4. **Plan for Testing**: Plan ahead for testing to make sure the data migrated is complete and accurate, by outlining test cases, methods, tools and technologies necessary for this step.

5. **Plan for Verification**: Plan for verification to ensure that migrated data is complete and accurate, outlining methods such as data integrity checks, reconciliation processes, and validation activities.

6. **Plan Post-Migration Activities**: Prepare for post-migration activities such as data cleanup, user training and system maintenance by outlining timelines and resources required.

10.2 CREATING A ROBUST DATA BACKUP & RECOVERY PLAN

Data backup and recovery plans provide organizations with a way of protecting critical data in case of disaster, including hardware failure, natural disasters or cyber attacks. Their goal is to quickly restore their critical files after any outages occur or failure occurs.

Create a Robust Data Backup and Recovery Plan:

A robust data backup and recovery plan must be developed as part of pre-migration activities. Here are some best practices for developing an inclusive backup and recovery strategy:

1. **Define Recovery Objectives**: The initial step in developing a data backup and recovery plan is defining recovery objectives, including both recovery point objectives (RPOs) and recovery time objectives (RTOs). RPOs represent how much data an organization is willing to lose during a disaster; RTOs specify how long an organization can go without access to its critical data.

2. **Recognizing Critical Data**: It is vitally important to identify critical data which should be prioritized based on RPO/RTO rules, such as financial records, customer data, business applications or any other essential documents required by an organization for

operation. Such critical data could include financial documents, customer details or business applications that support operations within an enterprise.

3. **Select Backup and Recovery Methods**: Backups come in different forms: full backups, incremental backups and differential backups are all options to consider when choosing an RPO and RTO ratio. Full backups require more storage space but can be restored more easily while incremental backups take less time but require more complex restoration steps.

4. **Test the Plan**: Once a backup and recovery plan has been developed, it is vitally important to conduct regular simulations of its functionality to make sure everything works as intended. Including simulation tests in regular meetings may ensure that it remains up-to-date and relevant.

Benefits of a Comprehensive Data Backup and Recovery Plan:

An effective data backup and recovery plan offers numerous advantages, such as:

1. **Minimizing Data Loss in Case of Disaster**: Such plans help safeguard critical information in the event of disaster, assuring it will be available when required.

2. **Reduce Downtime**: An effective backup and recovery plan can minimize downtime for an organization by providing them with the ability to continue operations even in the event of a catastrophe.

3. **Strengthen Security**: Having a backup and recovery plan in place can add another layer of protection for a business's sensitive information, by offering redundant storage solutions and making sure no data is lost or stolen.

10.3 ADOPTING INDUSTRY STANDARDS & FRAMEWORKS FOR PRE-MIGRATION ACTIVITIES

Organizations should enact industry standards and frameworks in their pre-migration activities for many reasons, the main one being that this can help ensure consistency and reliability during data migration processes. By following established frameworks, organizations can have confidence in knowing their pre-migration activities are being carried out efficiently and effectively.

Adopting industry standards and frameworks can assist organizations with recognizing potential issues and risks related to data migration processes. By following established best practices and frameworks, organizations can proactively address any issues that arise and take measures to minimize any potential threats before they become major problems.

Adopting industry standards and frameworks can also ensure organizations comply with regulations and standards relevant to them. Many industries face stringent regulatory requirements when it comes to data management and migration; by adhering to established frameworks, organizations can be sure that they are meeting all relevant governmental obligations.

Which Standards and Frameworks Should Organizations Adopt?

Which Standards and Frameworks Should Organizations Adopt for Pre-Migration Activities? Organizations may adopt various industry standards and frameworks during pre-migration activities. Some of the more frequently utilized frameworks include:

- **IT Infrastructure Library (ITIL):** This framework offers a comprehensive set of best practices for IT service management, including pre-migration activities.

- **Control Objectives for Information and Related Technology (COBIT):** This framework offers a comprehensive set of best practices for IT governance, including pre-migration activities.

- **Project Management Body of Knowledge (PMBOK)**: This framework offers a comprehensive collection of best practices in project management, including pre-migration activities.

- **International Organization for Standardization (ISO):** The ISO provides standards across a variety of industries, such as IT and data management.

Each framework has its own set of advantages and disadvantages, yet all can serve organizations looking to adopt best practices and standards for pre-migration activities. Ultimately, however, the choice will ultimately depend on each organization's individual requirements and needs.

Challenges in Adopting Industry Standards and Frameworks

While adopting industry standards and frameworks may provide organizations with many advantages, adopting them also presents unique challenges that must be met head on. Chief among them is making sure all stakeholders understand the chosen framework - this may require extensive training programs for team members who wish to effectively utilize it within their work environments.

An additional challenge lies in making sure the chosen framework satisfies the specific needs and requirements of an organization. Frameworks provide useful starting points, but must be adjusted according to each organization's circumstances.

Organizational resistance may impede any significant transformation. Adopting new standards and frameworks often requires changes to existing processes and procedures; some team members may oppose such modifications.

10.4 LEVERAGING AUTOMATION & TOOLS FOR PRE-MIGRATION ACTIVITIES

Pre-migration activities are critical in ensuring the success of a data migration project. They involve activities such as data profiling, data cleansing, and data validation. These activities can be time-consuming, labor-intensive, and error-prone if done manually. Leveraging automation and tools can help to reduce the time and effort required for these activities, while also increasing accuracy and reducing the risk of errors.

Different types of automation and tools available:

1. **Data Profiling Tools**: These tools enable organizations to assess the structure, content and quality of data within source systems. They can identify potential issues such as missing values, duplicates or inconsistencies within this data and recommend ways to address it.

2. **Data Cleansing Tools**: These tools assist in cleaning and standardizing data to ensure its consistency and accuracy, including eliminating duplicate records, correcting misspellings, and standardizing formats.

3. **Data Validation Tools**: These tools assist in verifying that data being migrated meets all necessary standards and is consistent with its target system. They can help identify any incorrect values within

data and offer solutions on how best to address any potential problems that might occur during migration.

Benefits of using automation and tools for pre-migration activities:

1. **Increased Efficiency**: Utilizing automation and tools can significantly decrease the time and effort required for pre-migration activities, helping organizations complete migration projects faster.

2. **Improved Accuracy**: Automation and tools can assist with minimizing risks associated with pre-migration activities, helping ensure that data being migrated is both precise and consistent.

3. **Reduced Risk**: Automation and tools can help minimize risks related to data loss, corruption and security breaches during migration processes.

4. **Consistency**: Automation and tools can assist with ensuring the data being migrated is consistent between various systems, making it simpler to integrate and use within its target environment.

10.5 ESTABLISHING A PRE-MIGRATION CENTER OF EXCELLENCE

Data migration is a complex process that requires extensive planning, resources and expertise. Establishing a pre-migration center of excellence can help ensure a smooth process - this team of experts develops and implements best practices to facilitate data migration while identifying any potential issues before developing strategies to address them and ensuring smooth execution during migration.

This section will introduce the concept and benefits of pre-migration centers of excellence, as well as their key components and practical advice for creating one in your organization.

Key Components of a Pre-Migration Center of Excellence:

1. **A Clear Mission and Goals**: The first step in creating a Pre-Migration Center of Excellence is outlining a clear Mission and Goals. A mission statement should clearly state its objectives while offering guidance to team activities. Additionally, specific, measurable, attainable, relevant, and time-bound goals should be set forth.

2. **Dedicated Team**: For optimal pre-migration center of excellence operations, it requires a dedicated team of specialists tasked with developing and implementing best practices for data migration. This group should possess diverse skill sets including project management, data analytics and technical acumen.

3. **Best Practices**: A pre-migration center of excellence should focus on creating and implementing best practices for data migration, using industry standards or frameworks as guides while still meeting the unique requirements of an organization.

4. **Documentation**: Documentation is an essential element of a pre-migration center of excellence. To ensure smooth migrations in future migration projects, the team should document all aspects of their migration process--such as project plans, risk analyses and testing results--including project plans, risk analyses and test results. Having this documentation will serve as a reference when making future migration plans and ensures smooth execution.

5. **Continuous Improvement**: Pre-migration centers of excellence should prioritize continuous improvement. Their teams should regularly review and update best practices and documentation to remain abreast of industry standards and trends.

Benefits of Establishing a Pre-Migration Center of Excellence:

1. **Reduced Risk**: Establishing a pre-migration center of excellence can help mitigate data loss, system downtime, and other potential issues that may arise during migration process. A team can quickly identify any potential issues and work on solutions before they become major problems.

2. **Increased Efficiency**: Establishing a pre-migration center of excellence can assist with increasing the efficiency of migration process by developing and implementing best practices, streamlining it and decreasing time and resource requirements during migration.

3. **Improved Data Quality**: Establishing a pre-migration center of excellence can help enhance the quality of the data being migrated. Through data profiling and cleansing activities before migration begins, teams can identify any data quality issues which may impede its success and address them immediately.

4. **Improved Communication**: Establishing a pre-migration center of excellence can assist with bettering communications among stakeholders. The team can act as a central point of contact for all migration-related activities and will keep all parties up-to-date on its progress.

10.6 ENSURING DATA SECURITY & PRIVACY IN PRE-MIGRATION ACTIVITIES

Data migration projects involve the transfer of sensitive information from one system to the other. When doing this, it is vitally important that data security and privacy are upheld so as to avoid unauthorized access or breaches that could lead to significant financial losses, regulatory fines and loss of customer trust - it is therefore imperative that pre-migration activities prioritize data protection to minimize any potential breaches in data protection or privacy.

Best Practices for Ensuring Data Security and Privacy during Pre-Migration Activities:

Conduct a Data Inventory:

Prior to initiating migration activities, it is crucial to conduct an inventory of the data that will be migrated. Doing this will enable you to identify sensitive elements which require special care during migration; once identified, these sensitive data elements can receive extra protection measures in order to guarantee their security during this phase.

Conduct a data classification:

Data classification involves categorizing information based on its level of sensitivity and protecting each element accordingly. By classifying their data, organizations can implement appropriate security measures and access controls to protect sensitive information.

Implement Access Controls:

Access controls are crucial in maintaining data security and privacy, and only authorized personnel should have access to sensitive data. Therefore, implementing them during pre-migration activities is key in order to safeguard any unauthorized access or manipulation of sensitive information.

Data encryption:

Data encryption refers to the process of converting information into a coded form that only authorized parties can decode, with encryption used during pre-migration activities to ensure sensitive information remains protected even if it falls into unintended hands. Using strong algorithms and keys ensures maximum data protection.

Conduct vulnerability assessments:

Vulnerability assessments allow organizations to detect any flaws in the system that attackers could exploit and then take necessary measures to address those risks. Vulnerability checks should be regularly carried out during pre-migration activities in order to ensure that your system is secure.

Conduct Penetration Testing:

Penetration testing involves simulating an attack against a system to identify vulnerabilities not revealed during vulnerability assessments. As part of its purpose, penetration testing helps organizations detect potential security breaches so they can take measures against them before any occur.

Implement Security Monitoring:

Utilizing monitoring tools helps security personnel detect any unusual activities that could indicate a security breach and take appropriate measures to stop data breaches from occurring. Monitoring tools provide alerts when suspicious activities arise and alert security staff of such activity so they can take appropriate actions quickly to mitigate data losses.

Train Employees:

Human error is one of the primary sources of data breaches, so training employees on data security and privacy best practices is of utmost importance. Through employee training sessions, it ensures they understand why data protection and privacy should be prioritized during pre-migration activities and know how to best handle sensitive information during this process.

10.7 ADDRESSING LEGAL & REGULATORY COMPLIANCE IN PRE-MIGRATION ACTIVITIES

Migration of data can be a complex and challenging project that demands careful planning and execution. One key part of this process involves meeting legal and regulatory compliance in pre-migration activities - failure to do so could result in costly penalties, legal disputes or irreparable damage to an organization's reputation; so it is crucial that legal and regulatory compliance issues be taken into consideration during pre-migration stage to mitigate these potential risks.

Legal and Regulatory Compliance:

Legal and regulatory compliance refers to adhering to laws, regulations, and policies established by governing bodies such as government agencies or industry associations. These laws and policies aim to protect individual rights while guaranteeing fair competition and maintaining public safety; with regards to data migration projects this can involve data privacy, security, intellectual property management and export controls among many others.

Challenges in Addressing Legal and Regulatory Compliance:

Adherence to legal and regulatory standards can present several unique obstacles during pre-migration activities, for many different reasons. First off, laws and regulations differ considerably depending on both industry type and geographic location. So it is essential for an organization to possess an in-depth knowledge of legal requirements applicable to its operating environment. Furthermore, compliance requirements may change over time, so their pre-migration plan must be adaptable enough to take account of this. Data migration involves moving large volumes of sensitive information across borders, leaving it exposed to cyber-attacks and unauthorized access. Maintaining data security while complying with legal and regulatory obligations can be a delicate balancing act.

Best Practices for Addressing Legal and Regulatory Compliance:

In order to ensure legal and regulatory compliance in pre-migration activities, organizations should adopt best practices which include these key items:

1. Conduct an exhaustive review of laws and regulations relevant to your industry and geographic location.

2. Gain an in-depth understanding of which data types fall under legal and regulatory compliance obligations.

3. Align data security and privacy policies with legal and regulatory compliance requirements.

4. Conduct regular security audits to detect potential vulnerabilities and risks.

5. Devise contingency plans in case of compliance violations or other legal issues.

6. Establish a communication plan to inform all stakeholders - employees and customers alike - of legal and regulatory compliance requirements as well as their responsibilities.

7. Make sure that all third-party vendors and service providers meet all legal and regulatory requirements, before creating a compliance monitoring program to evaluate efforts at compliance.

8. Create an evaluation framework so you can track compliance efforts more effectively.

9. Provide ongoing training and education to employees and stakeholders to ensure continued compliance.

10. Engage legal and regulatory specialists to ensure compliance efforts align with current laws and regulations.

10.8 SUMMARY

- For any successful data migration project, creating an extensive plan is key to its success.

- Establishing a reliable backup and recovery strategy helps mitigate risks related to data loss during migration.

- Adopting industry standards and frameworks for pre-migration activities helps create an organized process. Utilizing automation tools is also useful, reducing manual effort while decreasing errors.

- Establishing a pre-migration center of excellence helps provide a centralized approach to data migration activities. Ensuring data security and privacy during pre-migration activities helps maintain confidentiality while preventing data breaches.

- Pre-migration activities must consider legal and regulatory compliance to ensure compliance with relevant laws and regulations, while conducting an in-depth data analysis helps identify potential problems before creating plans to address them.

- Engaging stakeholders and communicating effectively are keys to ensuring a seamless migration.

- Carrying out extensive testing and validation helps verify the integrity and accuracy of migrated data.

10.9 TEST YOUR KNOWLEDGE

I. What is the first step in developing a comprehensive plan for data migration?

a) Identify key stakeholders
b) Select a migration tool
c) Develop a data backup plan
d) Establish a pre-migration center of excellence

II. What is the purpose of creating a robust data backup and recovery plan?

a) To ensure that data is not lost during migration
b) To speed up the migration process
c) To reduce costs associated with migration
d) To improve data quality

III. What is the benefit of adopting industry standards and frameworks for pre-migration activities?

a) Ensures legal and regulatory compliance
b) Improves data quality
c) Reduces migration costs
d) Increases migration speed

IV. What is the role of automation and tools in pre-migration activities?

a) To reduce errors and improve efficiency
b) To eliminate the need for human intervention
c) To increase migration speed
d) To reduce costs associated with migration

V. What is a pre-migration center of excellence?

a) A team responsible for overseeing all pre-migration activities
b) A set of industry standards and frameworks for pre-migration activities
c) A software tool for managing data migration
d) A data backup and recovery plan

VI. **Why is it important to ensure data security and privacy in pre-migration activities?**

a) To comply with legal and regulatory requirements
b) To improve data quality
c) To reduce migration costs
d) To increase migration speed

VII. What are some legal and regulatory compliance issues to consider in pre-migration activities?

a) Data privacy laws, intellectual property rights, and export control regulations
b) Data quality standards, migration speed, and cost reduction

c) Data backup and recovery plans, automation tools, and industry frameworks
d) Key stakeholders, project timelines, and budget constraints

VIII. **What is the benefit of establishing a pre-migration center of excellence?**

a) Improved coordination and oversight of pre-migration activities
b) Reduced migration costs
c) Increased migration speed
d) Improved data quality

IX. **What are some automation tools that can be used in pre-migration activities?**

a) Data profiling tools, data mapping tools, and data cleansing tools
b) Backup and recovery tools, migration planning tools, and risk management tools
c) Data analysis tools, migration testing tools, and migration reporting tools
d) Data encryption tools, data masking tools, and data archiving tools

X. **How can legal and regulatory compliance issues be addressed in pre-migration activities?**

a) By conducting a compliance review and developing a compliance plan
b) By focusing on data quality and data cleansing activities
c) By adopting industry frameworks and standards
d) By leveraging automation and tools

10.10 ANSWERS

I. Answer: a) Identify key stakeholders
II. Answer: a) To ensure that data is not lost during migration
III. Answer: a) Ensures legal and regulatory compliance

IV. Answer: a) To reduce errors and improve efficiency

V. Answer: a) A team responsible for overseeing all pre-migration activities

VI. Answer: a) To comply with legal and regulatory requirements

VII. Answer: a) Data privacy laws, intellectual property rights, and export control regulations

VIII. Answer: a) Improved coordination and oversight of pre-migration activities

IX. Answer: a) Data profiling tools, data mapping tools, and data cleansing tools

X. Answer: a) By conducting a compliance review and developing a compliance plan

11

11 CASE STUDIES AND EXAMPLES

Data migration is an ever-evolving discipline, and companies face the constant challenge of transitioning their information from one system to the next. Success of any data migration project depends upon various elements such as quality data, security procedures and measures, legal compliance issues and pre-migration planning. One of the best ways to understand and gain an understanding of these factors is through studying real-life examples and case studies of successful and unsuccessful data migration projects. This chapter will present several case studies and examples of data migration projects from across industries, highlighting key challenges and solutions implemented by organizations to ensure successful data migration projects. We hope that by sharing these case studies we can offer valuable insights and best practices for organizations embarking on their own data migration journeys.

11.1 SUCCESSFUL DATA MIGRATION OF A LARGE E-COMMERCE PLATFORM

Data migration can be an intimidating challenge for any organization, yet is even more crucial in an e-commerce platform that relies on timely and accurate data for operations. Businesses in today's competitive environment must migrate data faster with greater accuracy while minimizing downtime; any delay or error could potentially impact customer satisfaction as well as financial loss.

This section examines the successful data migration of a large e-commerce platform, detailing both its challenges during migration and its strategies to address them. This case study serves as a useful example for other organizations planning similar efforts.

Background

The e-commerce platform at hand was a leading retailer of fashion products with an expansive customer base and significant inventory. They had been using an outdated database system which wasn't scalable enough to handle growing transactions; consequently, they recognized they needed an upgrade so as to meet market trends and satisfy customer requirements.

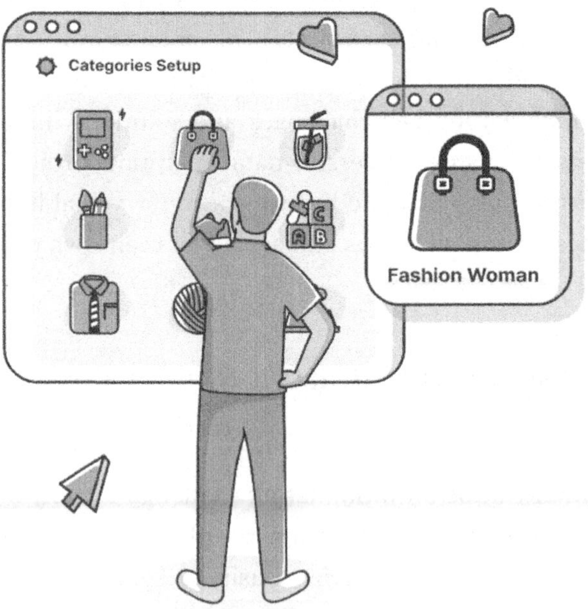

Challenge

The migration process presented several obstacles for our organization, including:

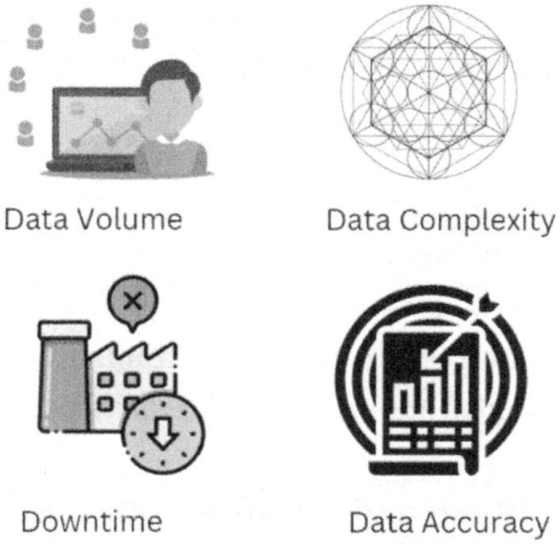

1. **Substantial Data Volumes**: The e-commerce platform had an enormous database containing customer details, orders, transactions and product info that needed to be moved quickly across to our new platform. Taking on such an enormous task presented an immense challenge that needed to be met swiftly in order for migration to occur successfully.

2. **Data Complexity**: The data was stored in different formats - structured, semi-structured and unstructured data were all present - making migration of such complex information an intricate task that required expert expertise and special tools.

3. **Downtime**: To avoid any loss of revenue or customers during migration, the platform had to remain operational throughout the migration process.

4. **Accuracy of Data**: Since customer information was key for personalized services and recommendations on this e-commerce platform. The strategy that underpinned all this work included remaining fully operational during migration as a precaution against losing any business opportunities due to any disruption caused by migration.

Strategy

1. **Planning**: In order to overcome these challenges, the organization adopted a systematic approach and planned the migration process carefully, outlining scope, timelines, resources needed and any potential risks.

2. **Data Profiling and Cleaning**: To ensure its accuracy, completeness, and consistency, the organization profiled and cleansed their data to ensure its accuracy, completeness, and consistency.

3. **Testing**: They conducted extensive tests of their new database system to ensure compatibility with its platform applications and data.

4. **Automation**: To migrate data efficiently and quickly, they utilized automation tools.

5. **Downtime Management**: They minimized downtime by scheduling the migration process during off-peak hours and creating a fallback plan in case any issues arose during migration process.

Results

Thanks to their strategic approach and thorough planning, this organization successfully executed its data migration with minimal impact to platform operations. Their new database system enabled them to scale operations while offering more personalized services for customers - leading to higher customer satisfaction ratings and revenue increases as a result of the migration.

11.2 CHALLENGES AND SOLUTIONS IN MIGRATING LEGACY SYSTEMS OF A FINANCIAL INSTITUTION

Migrating legacy systems of a financial institution is an arduous and time-consuming task that demands careful planning, execution, and management. Legacy systems may be outdated, complex, poorly documented, making them hard to understand or modify. Furthermore, legacy systems may be essential to operations and any disruption or downtime during migration may incur serious financial and reputational losses.

In this section, we will outline the challenges and solutions associated with migrating legacy systems from financial institutions. Additionally, we will detail best practices and tools which can help overcome such hurdles to ensure a successful migration.

Challenges in Migrating Legacy Systems of a Financial Institution

1. **Complexity and Lack of Documentation**: Legacy systems tend to be complex and poorly documented, making it hard for staff members to comprehend their functionality, data flows and dependencies. This lack of clarity regarding design, implementation and maintenance creates complications during migration processes as it makes identifying potential issues challenging.

2. **Inadequate Testing**: Testing is an essential element of any data migration project, and failing to conduct thorough tests could create unexpected issues during migration. Legacy systems typically lack comprehensive testing capabilities which leads to insufficiently testing migrated data which can result in loss or corruption of important files.

3. **Disruption to Business Operations**: Financial institutions operate in highly regulated environments, and any disruption or downtime during migration could lead to significant financial and reputational losses. Therefore, it is crucial that any migration plans be planned and executed with as little disruption to business operations as possible.

4. **Data Security and Privacy**: Financial institutions manage sensitive customer data that must remain private and secure, so any breaches in data protection could have serious legal and reputational repercussions. It is therefore essential that financial institutions ensure the protection and confidentiality of customer information during any migration process.

Solutions to Migrating Legacy Systems of a Financial Institution

1. **Comprehensive Planning**: Effective legacy system migration projects depend upon extensive planning. This requires creating a detailed project plan, outlining migration goals and objectives, identifying any risks or issues, allocating appropriate resources, etc.

2. **Documentation and Analysis**: Documenting and analyzing legacy systems is essential to fully comprehending their functionality, data flows and dependencies. This involves producing comprehensive system documentation as well as conducting in-depth analyses on its architecture to detect any potential issues or risks within them.

3. **Testing**: Thorough testing is necessary to detect and resolve potential issues during the migration process, such as testing migrated data, new system functionality, or legacy system integration with other systems.

4. **Business Continuity Planning**: Proper business continuity planning is essential to ensure minimal disruption during a migration. This should include creating an extensive contingency plan, identifying key processes and allocating adequate resources in order to minimize downtime or disruption during this transition process.

5. **Data Security and Privacy**: Security should always be of top importance during any migration project, from making sure data is encrypted during transit and rest, implementing access controls and monitoring, as well as complying with relevant regulations and standards.

11.3 MIGRATING HEALTHCARE DATA: LESSONS LEARNED AND BEST PRACTICES

Healthcare data migration processes must be efficient, accurate, and secure to provide high-quality patient care. Electronic Health Records (EHRs) have become the standard in modern medicine; with such vast amounts of patient information created daily by EHRs alone. As healthcare organizations generate this vast quantity of patient information daily, data migration must take place efficiently, accurately, and securely if they want their patient records not compromised in any way. Migrating healthcare data presents unique challenges which should be carefully considered prior to migrating projects such as this. In this section we explore these processes alongside insights learned from prior migration projects as well as lessons learned from previous migration projects undertaken successfully by healthcare organizations involved.

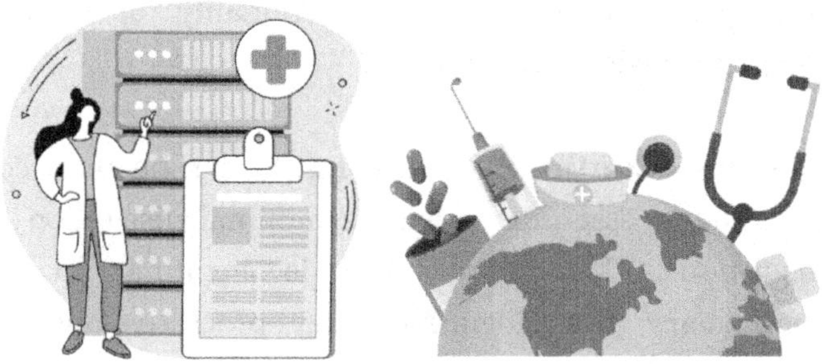

Challenges in Migrating Healthcare Data

1. **Complex Data Structures**: Healthcare data can be highly complex, generated from many disparate sources and systems such as EHRs, laboratory information systems, imaging systems and more. Each of these uses different formats, models and standards which makes data migration an arduous task.

2. **Data Security and Privacy**: Health data is highly sensitive, subject to rigorous regulatory compliance requirements, and its migration requires stringent security protocols and measures in order to maintain confidentiality, integrity, and availability of patient records.

3. **Data Quality**: For high-quality patient care, accurate and complete healthcare data are of utmost importance. While migrating data can present new quality issues such as inconsistencies, errors, and duplicates that can undermine care delivery; any migration-induced issues could potentially compromise data accuracy and integrity, which in turn compromise patient treatment quality.

4. **Business Continuity**: Healthcare organizations cannot allow any disruptions during data migration to affect patient care; as such, healthcare organizations must ensure their operations continue seamlessly during this process.

Best Practices in Migrating Healthcare Data

1. **Profiling and Analysis of Healthcare Data**: Before migrating their healthcare data, healthcare organizations should conduct a detailed examination and analysis. Data profiling helps organizations detect any inconsistencies, errors, or duplicates within their data which they need to address through migration strategies.

2. **Establish Data Governance**: Proper data governance policies and procedures can ensure the accuracy, completeness, and consistency of healthcare data. Healthcare organizations must establish clear procedures for data protection as part of

their data governance protocols to protect patient records properly.

3. **Create a Comprehensive Data Migration Plan**: Healthcare organizations should develop an inclusive migration plan which includes timelines, milestones, and specific tasks to be completed. Incorporation of risk mitigation strategies within this plan should help ensure business continuity.

4. **Train Staff and End-Users**: For a successful transition, both staff and end-users should receive adequate training on the new system and processes. Healthcare organizations must offer thorough instruction so staff and end-users are equipped to utilize the new system effectively.

Lessons Learned from Previous Migration Projects

1. **Engage End-Users Early**: To ensure their needs are addressed during migration projects, end-users should be involved from the outset of migration projects. Early involvement by end users can help identify potential issues early and ensure the new system satisfies their demands.

2. **Conduct Thorough Tests Before Implementation**: Healthcare organizations should perform exhaustive tests of any new system before going live, in order to identify any issues or errors which must be resolved before going live with it. Testing helps healthcare organizations detect any flaws which need addressing before going forward with implementation of their system.

3. **Monitor Data Quality**: Healthcare organizations must continuously assess data quality to ensure it remains accurate, complete, and consistent across patient records. Any discrepancies must be immediately identified and addressed so as not to compromise patient care.

4. **Address Legacy Data**: Healthcare organizations should address legacy data during their migration process. Converting or storing it separately so it will always be available when necessary is vital to ensuring patient data from historical periods remains readily available when needed.

11.4 DATA MIGRATION IN THE PUBLIC SECTOR: A CASE STUDY OF A GOVERNMENT AGENCY

Data migration is a complex task that requires meticulous planning, execution and monitoring. For public sector entities in particular, migration presents particular difficulties due to large volumes of data being moved between systems; protecting privacy while keeping services running uninterrupted for citizens is also paramount. In this section, a case study of one such government agency who successfully transitioned its data onto new platforms will be discussed highlighting both challenges encountered as well as best practices that could be utilized when similar cases arise.

Background

The government agency in question provided services such as tax collection, social security payments and healthcare to citizens. Over time, however, they had amassed large volumes of data across legacy systems that was difficult to access, manage or analyze. In order to increase efficiency and provide improved services to their constituents, they decided to migrate this data to a system capable of accommodating its volume and complexity. Challenges

Challenges

1. **Legacy systems**: The migration project did not come without challenges for the agency. It was necessary to address several key issues including legacy systems integration which required careful planning and execution before migrating the data over to a new system.

2. **Data Quality**: Due to its spread across multiple systems and variable quality, migrating the agency's data proved challenging.

3. **Data Security**: To protect both its own security as well as citizens' privacy during migration processes.

4. **Business Continuity**: To ensure business continuity during migration, services provided by the agency had to remain uninterrupted throughout its transition process, which required meticulous planning and coordination.

Best Practices

Even with all its challenges, the agency was able to migrate its data successfully onto its new system. Following are a few essential best practices which played a part in making sure this project succeeded:

1. **Planning**: To ensure a smooth migration process, the agency created an exhaustive plan that included an inventory of its data, identification of owners for specific collections, and setting a timeframe for its migration.

2. **Data Cleansing**: To ensure its data was accurate, complete, and consistent, this agency invested significant resources and time in data cleansing. 3. Testing: They extensively tested their new system to make sure that it met both functional and performance criteria.

3. **Training**: To ensure its staff were adept with using the new system effectively and were comfortable navigating it effectively. 5. Data Security: To protect its data during migration, including encryption and access controls.

4. **Communication**: The agency maintained regular dialogue with all of its stakeholders - including citizens - to keep them up-to-date with the migration process and address any concerns that they might have had about migration.

11.5 DATA MIGRATION IN THE RETAIL INDUSTRY: A CASE STUDY OF A GLOBAL RETAILER

Data migration refers to the act of moving data between systems or formats. For retailers in particular, data migration is essential in maintaining accurate inventory levels, managing customer information efficiently and improving overall business operations. In this case study we will look at one global retailer's data migration process as well as any challenges they experienced along with any solutions implemented to solve these challenges.

Company Background

This large multinational retailer had over 50 country presence and offered products including clothing, home goods and electronics to an established customer base. Due to their size and complexity, data migration proved an immense task.

Challenges Faced

One of the main challenges encountered by this retailer was migrating their massive amounts of data. Their multiple systems had differing formats for their data, which was difficult to consolidate into one centralized system. Furthermore, each store required its own inventory management system which needed to be integrated with this new one as part of the migration.

Assuring data accuracy was another challenge during migration. With such large volumes of information to transfer, there was always the risk of errors or inconsistencies which could have resulted in inaccurate inventory levels, customer information or other critical data sets being transferred - potentially leading to inaccurate inventory levels, customer profiles or other key records being displayed inaccurately.

Solutions Implemented

To meet these challenges, the retailer created a detailed data migration plan with an inventory of data sources as well as timeline and budget considerations for this project. In addition, a dedicated team of data migration specialists was appointed to oversee this process.

Solutions Implemented

To address issues surrounding data accuracy, the retailer established data cleansing procedures. This included eliminating duplicate records, standardizing formats and correcting any errors as they arose, as well as creating quality metrics to ensure the data migration met standards.

For better inventory management systems integration, the retailer implemented a data integration platform. Using this tool they were able to extract data from different systems, format it and load it into their new system.

Results

The retailer found their data migration project to be an outstanding success, consolidating all of their information into one system to gain better insights into their operations, increase inventory management efficiency, and create a personalized customer experience. They completed it quickly and within budget with minimal disruption to daily business operations.

11.6 MIGRATING DATA TO CLOUD: A CASE STUDY OF A SOFTWARE COMPANY

As organizations adopt cloud-based solutions, data migration to the cloud has become ever more critical. When migrating software company data to the cloud requires careful consideration for specific challenges and considerations that come up during migration process. This case study provides insights into one such successful data migration project undertaken by one software company to cloud; including key challenges faced and best practices implemented to ensure seamless transfer.

Current State:

This software company had its data stored on-premise, making scaling, maintenance, and security difficult. Furthermore, their information was spread out among different locations making access difficult. As a result, they recognized the need to move to the cloud in order to reduce complexity, increase agility, and achieve cost savings.

Challenges:

Migrating data to the cloud presented several challenges for our software company, including:

1. **Data Compatibility**: Because their data was stored in various formats, making compatibility with the cloud challenging.

2. **Data Security**: The company was concerned about any data breaches during migration, as well as ensuring their information would remain adequately secure on cloud storage platforms.

3. **Downtime**: Migration processes will necessitate some degree of downtime that would negatively impact business operations and customers alike.

4. **Data Quality**: The company was concerned about the quality of the migrated data, ensuring that the data remained consistent and accurate.

Solutions:

To meet this challenge, the software company employed several best practices to ensure an easy transition:

1. **Profiling and Cleansing**: To ensure data quality and compatibility with cloud storage environments, the company conducted an intensive data profiling and cleansing process.

2. **Cloud Security**: To ensure the highest levels of data security and protection, this company collaborated closely with their cloud provider in implementing encryption and access controls.

3. **Data Replication**: Utilizing data replication technology allowed them to minimize downtime for an easier transition into the cloud environment.

4. **Testing**: To ensure data migration was accurate and consistent. 5. Training: Employees received training on their new cloud-based system before it went live. Results: Once migration had taken place, results were apparent within weeks.

Results:

A software company's transition to the cloud was a triumphant one, achieving these outcomes:

1. **Reduced Complexity**: By streamlining its data management processes, they were able to improve efficiency while cutting costs.

2. **Increased Agility**: The company was able to easily scale its operations as necessary, improving its ability to adapt quickly to changing business needs.

3. **Enhance Security**: With robust security measures protecting against data breaches in place in the cloud environment, company data was safer.

4. **Improved Data Qual**ity: With its data more consistent and accurate in the cloud, the company was better positioned to make informed business decisions.

11.7 CHALLENGES IN MIGRATING DATA TO A NEW ERP SYSTEM

Migrating data to a new enterprise resource planning (ERP) system can be a complex and challenging project. Success of such migration depends upon several factors including quality of data, compatibility between old and new systems, as well as expertise of team responsible for migration.

In this section, we will look at some of the challenges organizations may encounter when migrating their data to an ERP system and highlight some best practices they can adopt to ensure a smooth migration.

Challenges in Migrating Data to a New ERP System:

1. **Data Quality**: One of the key challenges associated with migrating data to a new ERP system is data quality issues such as missing or duplicate entries that can cause significant disruptions in its use in its new form. Therefore, before initiating migration proceedings it is vital to ensure all records are complete, accurate, and up-to-date before beginning this journey.

2. **Data Compatibility**: Another challenge associated with migrating data to a new ERP system lies in its compatibility. Data formats, field names and structures may differ between systems which further complicates migration efforts and mapping data from old systems to new.

3. **Business Process Changes**: Transitioning to a new ERP system may involve making modifications to your organization's business processes, which can present a considerable challenge as your employees adapt to an entirely different way of working. It is vital that business process owners are involved with this migration process so that their requirements are fulfilled by the new system.

4. **System Configuration**: Adopting an ERP system can be a complex process. In order for it to meet an organization's individual requirements and operate efficiently and effectively, correct configuration must be executed correctly in order to achieve successful implementation.

5. **Data Security**: Data security should always be considered when moving data to a new ERP system. Protecting it from unapproved access or breaches during migration requires setting up appropriate controls and protocols in order to guarantee its protection.

Best Practices in Migrating Data to a New ERP System:

1. Plan in Advance: To ensure the success of any data migration project, proper planning is key to its completion. Creating a comprehensive plan outlining all activities needed during migration is imperative for its successful execution.

2. Engage Key Stakeholders: Engaging key stakeholders such as business process owners, IT staff and end users in the migration process is imperative in order to ensure that any new system will meet all organization requirements.

3. Careful Testing: Testing is an integral component of data migration. Involve all parties involved and conduct comprehensive evaluation of your new system prior to migrating any data - this will identify any issues or potential concerns which need addressing before beginning the migration process.

4. Train End-Users: Effective end-user training is crucial to the success of any data migration project. Providing thorough instruction ensures that end users become acquainted with their new system, and can utilize it efficiently.

5. Implement Data Governance: Data governance is critical in order to guarantee accurate and complete information.

Therefore, it's crucial that appropriate policies and procedures be set in place in order to properly manage this resource.

11.8 A CASE STUDY ON MIGRATING DATA ACROSS BORDERS: LEGAL AND REGULATORY CONSIDERATIONS

Data migration across borders has become an increasingly prevalent practice for businesses today, though with various legal and regulatory frameworks across nations it can be challenging to ensure compliance and security during this process. In this case study we will look at one multinational corporation's experiences migrating its data across borders while taking note of any legal or regulatory considerations that needed to be addressed during their migration journey.

Background:

A corporation involved with technology needed to transfer a large volume of customer and financial data between subsidiaries in two different countries for purposes of centralizing it, improving efficiency, as well as meeting various legal and regulatory obligations.

Challenges:

Upon migrating such sensitive information as customer data, financial records, and intellectual property to their main headquarters located elsewhere, their primary concern became compliance with legal regulations as well as complying with other governmental rules and requirements.

Migration process was fraught with numerous hurdles, such as:

1. **Data Privacy Laws**: One of the subsidiary countries had stringent data privacy regulations which required specific

consent before data could be transferred out of country - meaning the corporation needed to gather consents from all involved before moving forward with migration process.

2. **Data Protection Laws**: The corporation had to ensure that appropriate measures were taken during migration of data between locations as each nation has individual regulations on data protection that mandate specific safeguards be put into place to safeguard personal information. This required them to take the necessary safeguards when migrating the data across borders.

3. **Cross-Border Data Transfer Restrictions**: As some subsidiary countries had restrictions regarding cross-border data transfers that required specific conditions be met before proceeding with migration projects, corporations had to ensure all necessary conditions had been fulfilled before initiating migration operations.

4. **Time Zone Differences**: Due to being located in different time zones, it was challenging for us to coordinate the migration process and ensure no interruption was caused in business operations.

Solutions:

1. **Legal Advice**: In response to these challenges, the corporation sought legal advice from legal experts both within its subsidiaries and headquarters countries in order to meet all legal and regulatory obligations.

2. **Consent Forms**: The corporation created custom-tailored consent forms that were compliant with data privacy

regulations in each subsidiary country, and obtained individual consents from all affected parties.

3. **Data Security Measures**: To safeguard data during its migration process, the corporation implemented appropriate data protection measures including encryption and access controls.

4. **Compliance Audits**: The corporation conducted compliance checks to ensure all requirements for cross-border data transfers had been fulfilled and obtained any necessary authorizations from relevant authorities.

5. **Project Management**: To minimize business disruption during migration, the corporation enlisted the services of a dedicated project team, with regular communication between subsidiaries and headquarters teams to coordinate.

11.9 MIGRATING DATA IN THE MANUFACTURING INDUSTRY: A CASE STUDY OF A LEADING MANUFACTURER

Manufacturing industries are among the most data-heavy sectors. From supply chain management and production operations to customer relations and customer data storage needs, data plays an essential role in every aspect of manufacturing. Therefore, data migration is an essential process for manufacturers looking to keep their operations running smoothly and efficiently; here, we explore both challenges and best practices of data migration for manufacturing businesses using a case study from one leading manufacturer.

Background:

This global manufacturer was an industry-leader in producing industrial equipment and machinery, operating across multiple countries around the globe and possessing an expansive product portfolio. Because of the complexity and volume of data involved, they had previously relied upon an outdated legacy system for managing it all; as they expanded further however, their legacy system became inadequate at meeting their changing needs; consequently they decided to migrate their data onto a modern system capable of meeting both current and future requirements.

Challenges in Migrating Data in Manufacturing Industries:

Migrating data across manufacturing industries presents unique hurdles which must be surmounted for successful outcomes. Below are a few challenges the manufacturer encountered during their data migration process:

1. **Data Volume and Complexity**: As the manufacturer had an overwhelming amount of information stored on its legacy system, its migration proved challenging. Furthermore, its data was diverse with multiple formats, types, and structures which necessitated careful planning so as to transfer all its details accurately into its new platform.

2. **System Integration**: As part of their legacy system integration strategy, this manufacturer had multiple applications and systems intertwined. Migrating their data without disrupting these integrations was essential in maintaining smooth operations.

3. **Downtime:** As any extended downtime during migration would compromise production operations and cause financial

losses, minimizing downtime while guaranteeing accurate data transfer was an enormous challenge for this manufacturer.

4. **Data Security**: To safeguard their operations and avoid data leakages that could threaten its operations, manufacturers needed to ensure all transferred data were securely transmitted into their new system.

Best Practices:

To address these challenges and ensure a successful data migration, the manufacturer implemented several best practices:

1. **Planning and Prep**: The manufacturer invested significant effort in planning and prepping for their migration process. This involved defining its scope, developing a migration strategy, setting a timeline, and allocating resources necessary for their move.

2. **Data Profiling and Cleansing**: The manufacturer conducted data profiling to understand the volume, quality and complexity of its data. Additionally, duplicate, incomplete or irrelevant records that could hinder accurate migration were removed through data cleansing processes.

3. **System Testing**: Prior to initiating migration, the manufacturer conducted extensive system tests on their new system in order to ensure compatibility with existing systems and applications, thus helping minimize downtime and prevent disruption during this process.

4. **Data Encryption and Security**: To protect their customers' sensitive data when transitioning to their new system, this manufacturer implemented data encryption as a first line of

defense against potential breaches in security. Furthermore, access controls were implemented to restrict unauthorized access to this data.

11.10 MIGRATING DATA FOR BUSINESS CONTINUITY: A CASE STUDY OF A LARGE SERVICE PROVIDER

Data migration is a complex and delicate process that must be planned carefully in order to succeed. It involves moving information between systems with the aim of increasing performance, scalability or functionality; but migration requires more than simply moving files from A to B - it necessitates an in-depth knowledge of both systems involved as well as of any potential business impacts from such changes.

In this section, we will focus on one specific use case of data migration: migrating for business continuity. Business continuity refers to an organization's ability to continue operations even when faced with an unplanned disruption, such as natural disasters, cyberattacks or any other threats to its operations.

Migrating data for business continuity purposes is a key strategy to ensuring an organization can quickly recover from an unexpected disruption. We will explore an example of an organization which successfully utilized migration for this purpose by exploring a case study of one large service provider who conducted migration for this purpose

Section 1: Background and Context:

In this section, we will present some background information about the service provider and context for data migration.

Service provider at hand is a global telecommunications giant offering internet, cable TV and phone services in multiple countries globally to millions of customers.

Data migration began as part of the company's business continuity plan. Due to previous disruptions like natural disasters and cyberattacks compromising their ability to provide services to customers, this company recognized the necessity of having a robust plan in place so it could continue providing services even during unexpected disruptions.

As part of its business continuity plan, the company made the decision to migrate its data onto a cloud-based solution in order to access their files from anywhere - guaranteeing their ability to continue operations if their physical offices became inaccessible.

Section 2: Challenges and Solutions

In this section, we will review some of the challenges encountered during data migration as well as possible solutions implemented to overcome them.

Challenge 1: Data Volume

The primary challenge faced by this company was migrating an immense volume of data across its various locations, which consisted of both on-premises data centers as well as remote locations.

Solution 1: Data Profiling and Cleansing

In response to this challenge, the company undertook an intensive data profiling and cleaning exercise. This involved reviewing their database in detail in order to detect duplicate records, inconsistencies or errors before working to cleanse it by removing unnecessary information.

Challenge 2: Data Security

The company faced another significant obstacle during the migration process - protecting its sensitive customer data without compromise during migration. This task had to be carried out carefully to avoid any mishaps.

Solution 2: Data Encryption and Secure Connectivity

To address this challenge, the company employed data encryption and secure connectivity protocols to protect data during migration. In addition, strict access controls were put into place so only authorized personnel had access to it.

Challenge 3: Data Integrity

The third challenge the company faced during their migration process was maintaining data integrity during migration. To do this effectively and without data loss or corruption occurring along the way.

Solution 3: Data Validation and Testing

To address this challenge, the company implemented an intensive data validation and testing process that involved checking every stage of data migration to ensure its accuracy and completeness as well as conducting extensive tests to detect any data losses or corruptions that might have taken place.

Solution: An application with an accurate data migration plan to ensure that data was transferred safely and accurately was created, and then manually and automated testing conducted to confirm its accuracy.

One of the major obstacles faced by the team was managing such an immense volume of data to be transferred. To address this challenge, they relied on automated solutions and utilized tools capable of quickly moving large volumes.

Additionally, the team put emphasis on data security and privacy throughout the migration process. They implemented measures to prevent data breaches while protecting sensitive information; these included encryption technologies, access controls systems and other security protocols.

Overall, the migration was successful and allowed the service provider to carry on operations without significant interruptions. Thanks to careful planning and execution by our team, data was transferred safely and accurately so as to maintain business continuity at both locations.

Lessons Learned:

- To ensure business continuity, data migration requires careful planning. Automation can be an efficient means of migrating large volumes of information quickly.

- Maintaining data security and privacy during any migration process must always be of top priority.

- Both manual and automated testing should be employed to test accuracy of migrated data.

- Involving a professional migration specialist for planning, execution and follow-up will help prevent disruptions while maintaining business continuity.

11.11 SUMMARY

- Case studies provide real-life examples of data migration projects and their challenges.

- Each case study highlights the unique aspects of each industry - healthcare, finance, retail and manufacturing among them.

- For successful data migration to occur, careful planning and execution must occur, with emphasis placed on data profiling, cleansing and backup/recovery processes as well as adhering to industry standards, using automation tools for data migration purposes, creating a pre-migration center of excellence, assuring data security/privacy as well as meeting legal/regulatory compliance issues.
- Each case study details how these pre-migration best practices were put into action and their impact on the success of a project.
- These case studies also emphasize the necessity of communication, collaboration, and teamwork among all parties involved in data migration processes.
- Data migration presents many unique challenges, including legacy systems, data quality issues, integration with existing systems and regulatory compliance requirements.
- Case studies offer insight into how these issues were overcome.
- Best practices identified in case studies include developing a comprehensive project plan, conducting rigorous testing and validation processes, and having a backup and recovery plan in place.
- These case studies also demonstrate the advantages of cloud migration, including increased scalability, flexibility, and cost savings. The lessons from these case studies can inform future data migration projects while helping organizations avoid common pitfalls and challenges.

11.12 TEST YOUR KNOWLEDGE

I. Which industry is the case study "Successful Data Migration of a Large E-commerce Platform" based on?

a) Healthcare
b) Retail
c) Finance
d) E-commerce

II. Which financial institution's legacy system is discussed in the case study "Challenges and Solutions in Migrating Legacy Systems of a Financial Institution"?

a) Citibank
b) Bank of America
c) JP Morgan Chase
d) Wells Fargo

III. What is the focus of the case study "Migrating Healthcare Data: Lessons Learned and Best Practices"?

a) Challenges of migrating data to the cloud
b) Migrating data of a global retailer
c) Lessons learned in migrating healthcare data
d) Migrating data of a government agency

IV. Which government agency is the case study "Data Migration in the Public Sector: A Case Study of a Government Agency" based on?

a) FBI
b) IRS
c) NASA
d) CDC

V. **Which global retailer is the case study "Data Migration in the Retail Industry: A Case Study of a Global Retailer" based on?**

a) Walmart
b) Amazon
c) Target
d) Alibaba

VI. **What is the focus of the case study "Migrating Data to Cloud: A Case Study of a Software Company"?**

a) Migrating healthcare data
b) Migrating legacy systems of a financial institution
c) Migrating data to the cloud
d) Migrating data of a government agency

VII. **Which ERP system is the case study "Challenges in Migrating Data to a New ERP System" based on?**

a) SAP
b) Oracle
c) Microsoft Dynamics
d) Infor

VIII. **What is the focus of the case study "A Case Study on Migrating Data Across Borders: Legal and Regulatory Considerations"?**

a) Challenges in migrating data to the cloud
b) Migrating data of a global retailer
c) Legal and regulatory considerations in cross-border data migration
d) Migrating healthcare data

IX. **Which manufacturing company is the case study "Migrating Data in the Manufacturing Industry: A Case Study of a Leading Manufacturer" based on?**

a) General Electric
b) Ford
c) Toyota
d) Siemens

X. **What is the focus of the case study "Migrating Data for Business Continuity: A Case Study of a Large Service Provider"?**

a) Migrating data to the cloud
b) Migrating legacy systems of a financial institution
c) Migrating data for business continuity
d) Migrating data of a government agency

11.13 ANSWERS

I. Answer: d) E-commerce
II. Answer: c) JP Morgan Chase
III. Answer: c) Lessons learned in migrating healthcare data
IV. Answer: b) IRS
V. Answer: a) Walmart
VI. Answer: c) Migrating data to the cloud
VII. Answer: b) Oracle
VIII. Answer: c) Legal and regulatory considerations in cross-border data migration
IX. Answer: d) Siemens
X. Answer: c) Migrating data for business continuity

Appendix A

A. COMMONLY USED DATA MIGRATION TOOLS & SOFTWARES

1. **Amazon Web Services (AWS) Database Migration Service:** A cloud-based tool from AWS that facilitates secure and seamless database migrations to AWS.

2. **Carbonite Migrate**: A data migration software offering automatic server and data migration across physical, virtual, and cloud environments.

3. **CloudEndure Migration**: A cloud-based migration tool that automates applications' transitions to the cloud, while minimizing downtime and data loss.

4. **DMS** iQ: An automated tool designed to uncover and assess database workloads in order to simplify the planning and execution of database migrations.

5. **IBM InfoSphere DataStage**: A robust data integration and migration tool with high scalability and dependability for large-scale data migration projects.

6. **Microsoft Azure Database Migration Service**: A cloud-based tool that makes the migration of on-premises databases into Azure easy.

7. **Oracle GoldenGate**: Data replication and integration software that enables real-time data migration between databases.

8. **Pentaho Data Integration**: An open-source data integration and migration tool with powerful ETL capabilities for complex data migration projects.

9. **SAP Data Services**: A comprehensive data integration and migration suite featuring advanced data quality and profiling features.

10. **Talend Data Integration**: An open-source data integration and migration tool that supports various data sources and formats.

These tools offer various features and capabilities to assist with data migration, such as automation, analysis, replication, and integration. It is essential to carefully assess and select the suitable tool based on your specific needs and requirements for the migration project.

Appendix B

B. COMMONLY USED DATA PROFILING TOOLS & SOFTWARES

1. **Talend Data Profiler**: This open-source tool enables users to discover, profile and analyze data from various sources while providing data quality metrics and reports.

2. **IBM InfoSphere Information Analyzer**: A powerful data profiling solution, capable of analyzing large volumes of information from various sources to produce data quality metrics, data lineage information and impact analyses.

3. **Trifacta**: This tool enables users to visually explore and profile large datasets, as well as provides various data quality metrics like completeness, accuracy, and consistency.

4. **DataCleaner:** An open-source tool offering advanced data profiling and analysis features such as quality metrics, matching algorithms, and standardization.

5. **Talend Open Studio**: A free and open-source ETL tool that includes data profiling and analysis capabilities as well as data quality metrics and reports.

6. **Informatica Data Quality**: A data profiling and analysis solution that offers a range of data quality metrics such as completeness, accuracy, consistency, and timeliness.

7. **Oracle Data Profiling**: A data profiling and analysis tool that offers various data quality metrics such as completeness, accuracy, consistency, and uniqueness.

8. **SAS Data Quality**: A powerful data profiling and analysis solution that offers a range of data quality metrics such as completeness, accuracy, consistency, and timeliness.

9. **Talend Data Preparation**: A data preparation solution that offers data profiling and analysis capabilities as well as a range of data quality metrics like completeness, accuracy, and consistency.

10. **Microsoft SQL Server Data Quality Services**: A data profiling and analysis tool that offers a range of data quality metrics such as completeness, accuracy, consistency, and timeliness.

Appendix C

C. COMMONLY USED DATA CLEANSING TOOLS & SOFTWARE

1. **OpenRefine**: An innovative open source tool for data cleaning, transformation and reconciliation. It supports various data formats as well as providing an intuitive interface to explore and edit large datasets.

2. **Talend Data Preparation**: A straightforward data cleaning tool that enables users to visually explore, clean and enrich data. It offers features like data profiling, quality assurance and enrichment.

3. **Trifacta**: A cloud-based data cleaning platform that enables users to visually explore, clean and transform large datasets. It offers an intuitive user interface as well as powerful machine learning algorithms for automation of the cleaning process.

4. **Data Ladder**: A powerful data cleansing platform offering features such as data profiling, deduplication and standardization. It supports various data formats and utilizes advanced matching and fuzzy clustering algorithms to enhance accuracy.

5. **Data Wrangler**: A free, open source data cleansing tool with an intuitive user interface to transform and clean data. It

supports various data cleansing operations such as splitting, merging, and filtering.

6. **IBM InfoSphere DataStage**: An enterprise-class data cleansing tool offering features such as data profiling, quality assurance and integration. It boasts powerful data transformation and matching capabilities with support for various data formats.

7. **Informatica PowerCenter:** A robust data integration and cleansing platform, offering features such as data profiling, quality assurance, and transformation. It boasts advanced data matching and standardization capabilities as well as support for various data formats.

8. **Alteryx**: An impressive data integration and cleansing tool that offers features such as data profiling, quality assurance, and transformation. With its user-friendly interface and advanced machine learning algorithms to automate the cleaning process.

9. **WinPure**: A comprehensive data cleansing tool offering features such as data profiling, deduplication and standardization. It supports various data formats and has advanced matching and fuzzy clustering algorithms.

10. **DataMatch Enterprise**: This enterprise-class data cleansing tool offers features like data profiling, quality assurance and standardization. It boasts advanced matching and deduplication capabilities as well as support for various data formats.

Appendix D

D. GLOSSARY

Here are some of the key terms and definitions included in the glossary:

A

- **Acceptance Testing**: Evaluating data migration results to verify they meet specified requirements and produce desired outcomes.
- **Automation**: Utilizing technology and tools to carry out tasks and processes with minimal human involvement.
- **Archive**: A repository of historical records or documents that are preserved for future reference.

B

- **Business Continuity**: The ability of a business to maintain essential operations during and after a disaster, such as a data center failure, cyber attack or natural calamity.
- **Backup**: The process of creating a copy of data to be used in the event of data loss or corruption.

C

- **Compliance**: Adhering to legal, regulatory or industry-specific requirements regarding data management, storage and transfer.

- **Cloud Computing**: A model for providing ubiquitous, on-demand access to a shared pool of computing resources that can be quickly provisioned and released with minimal management effort.
- **Cutover**: When the new system replaces the old one and transition to it is complete.
- **Center of Excellence (CoE)**: A team or department within an organization responsible for developing and sharing best practices and expertise in a given area.

D

- **Data Mapping**: Constructing a link between data elements in one system and those in the target one.
- **Data Migration**: Transferring information from one system to another, such as from an old one to a new one.
- **Data Quality**: Ensuring the accuracy, completeness, and consistency of data which can affect the success of a data migration project.
- **Data Privacy**: Protecting personal or sensitive information from unauthorized access, use or disclosure.
- **Data Security**: Ensure the protection of personal data against unauthorized access, use or disclosure.

E

- **Extraction**: The process of retrieving data from a source system, such as a database or a file system.

- **ETL (Extract, Transform, Load):** A process that extracts data from source systems, transforms it to fit the target system's requirements, and loads it into the target system.
- **ERP (Enterprise Resource Planning)**: A software system used to manage and automate business processes, such as accounting, inventory, and customer relationship management.

F

- **File Transfer**: The process of moving files from one system to another, which is often used in data migration.
- **Failover**: The process of switching to a redundant or standby system in the event of a failure or outage.
- **Framework**: A set of guidelines, standards, and practices used to structure, plan, and execute a project or process.

G

- **Governance**: The establishment of policies, procedures, and guidelines to ensure that data is managed effectively and efficiently.
- **Greenfield**: A term used to describe a completely new system or environment, without any pre-existing data.

H

- **High Availability**: The ability of a system or application to remain operational and accessible, even in the event of component failures or outages.
- **Heterogeneous System**: A system that consists of different hardware or software components.

I

- **Integration**: Consolidating data from different systems or applications.
- **Implementation**: Installing and configuring a new system or application.
- **Industry Standards**: These are guidelines and practices established by industry associations or organizations to guarantee quality, consistency, and interoperability in a given area.
- **Integration Testing**: A type of software testing that examines individual modules or components as a group to guarantee they operate together correctly. Integration testing plays an integral role in pre-migration activities.

J

- **Job scheduling**: The process of defining and scheduling jobs or tasks to run at specific times or intervals. In the context of data migration, job scheduling is often used to schedule data migrations and other related tasks.

K

- **Key performance indicators (KPIs):** Metrics used to evaluate the success or progress of a project or process. KPIs can be used to measure the success of pre-migration activities and to ensure that data migration projects are meeting their objectives.

L

- **Legacy systems**: Older or outdated computer systems or software that are no longer being actively developed or maintained. Legacy systems can pose significant challenges in data migration projects and require careful planning and management.

M

- **Mapping**: The process of defining relationships between data elements in the source and target systems. Mapping is an important part of pre-migration activities and helps ensure that data is accurately migrated from one system to another.

N

- **Non-functional requirements**: Requirements that define how a system should perform rather than what it should do. Non-functional requirements, such as performance, scalability, and availability, are an important consideration in pre-migration activities.

O

- **Objectives**: The goals or aims of a project or process. Objectives are an important consideration in pre-migration activities and help ensure that data migration projects are successful.

P

- **Project management**: The process of planning, organizing, and managing resources to achieve specific goals or objectives. Project management is an important part of pre-migration activities and helps ensure that data migration projects are completed on time and within budget.

- **Process mapping**: The process of documenting and analyzing a process to identify areas for improvement. Process mapping is an important part of pre-migration activities and can help identify inefficiencies in existing processes that can be addressed during data migration.

- **Proof of concept**: A demonstration of a concept or idea to show that it is feasible or practical. Proof of concept is often used in pre-migration activities to test migration processes and ensure that they will work as expected.

Q

- **Quality assurance**: The process of ensuring that a product or service meets specified quality standards. Quality assurance is an important part of pre-migration activities and helps ensure that data migration projects are completed successfully and meet the needs of stakeholders.

- **Query**: A request for specific data from a database or system.

- **Quick Wins**: Small and achievable tasks or goals that can be accomplished quickly and provide immediate benefits to the migration project.

R

- **RACI Matrix**: A tool used to define and assign responsibilities and roles to team members involved in a migration project.

- **Recovery Point Objective (RPO)**: The maximum acceptable amount of data loss in the event of a disaster or system failure.

- **Recovery Time Objective (RTO)**: The maximum acceptable amount of downtime before system recovery is achieved.

- **Risk Assessment**: A process of identifying and evaluating potential risks and vulnerabilities in the migration process.

S

- **Scope**: The boundaries and limitations of a data migration project, including the data, systems, and processes involved.
- **Source System**: The system or application from which data is being migrated.
- **Stakeholder**: An individual or group with an interest or stake in the migration project.
- **System Integration Testing (SIT):** Testing the integration of the migrated data with the target system to ensure compatibility and functionality.

T

- **Target System**: The system or application to which data is being migrated.
- **Technical Requirements**: The technical specifications and requirements for the migration process, such as hardware and software requirements.
- **Test Plan:** A document outlining the testing strategy, methodology, and objectives for the migration project.
- **Transformation**: The process of converting data from one format to another during migration.

U

- **User Acceptance Testing (UAT):** Testing the migrated data with real-world scenarios and use cases to ensure that it meets the end-user requirements.
- **User Requirements:** The specific requirements and needs of end-users for the migrated data.
- **Utility Program**: A software tool or program used to facilitate or automate specific tasks in the migration process.

V

- **Validation**: The process of verifying and ensuring that the migrated data is accurate, complete, and consistent.
- **Version Control**: The process of managing and tracking different versions of data and software during the migration process.
- **Volume Testing:** Testing the migrated data with large data sets to ensure scalability and performance.

W

- **Workflow**: The sequence of steps and processes involved in the migration project, including data extraction, transformation, and loading.
- **Workload**: The amount of data and processing required for the migration project.
- **Write-back**: The process of updating or writing data back to the source system after migration.

X

- **XML**: A markup language used to describe and transfer data between systems during migration.

Y

- **Yardstick:** A tool used to measure the success and progress of the migration project.
- **Yield**: The amount of usable data successfully migrated in relation to the amount of data processed.
- **Yottabyte**: A unit of data storage equal to one septillion bytes.

Z

- **Zero Downtime**: The goal of minimizing or eliminating downtime during the migration process.
- **Zip File:** A compressed file format used to package and transfer data during migration.
- **Zone**: A specific area or subset of data being migrated, such as a geographic region or business unit.

A HUMBLE REQUEST FOR FEEDBACK!

Dear Readers,

Hello, I hope this message finds you well. I am reaching out to ask for your opinion on this book "**The Key to Successful Data Migration**", which provides a comprehensive guide for individuals seeking to master data migration techniques.

As an author, my goal is to craft high-quality content that's both informative and accessible. This book serves as a great resource for data migration professionals of any level - whether they are novices or experienced experts.

Thank you so much if you could take the time to leave a review of my book. Your insight and opinion help me refine my writing, thus creating even better content for future readers. Thank you again for being here, and I hope that you found my book informative and useful.

Best regards,

Rajender Kumar

Made in the USA
Monee, IL
04 November 2026

67378709R00184